NG-RAN and 5G-NR

NG-RAN and 5G-NR

5G Radio Access Network and Radio Interface

Frédéric Launay

iSTE

WILEY

First published 2021 in Great Britain and the United States by ISTE Ltd and John Wiley & Sons, Inc.

ISTE Ltd
27-37 St George's Road
London SW19 4EU
UK

www.iste.co.uk

John Wiley & Sons, Inc.
111 River Street
Hoboken, NJ 07030
USA

www.wiley.com

Library of Congress Control Number: 2021935849

British Library Cataloguing-in-Publication Data
A CIP record for this book is available from the British Library
ISBN 978-1-78630-628-9

MIX
Paper from responsible sources
FSC www.fsc.org FSC® C013604

Contents

Preface

This book aims to describe the deployment of 5G-NSA and 5G-SA.

For the 5G-NSA mode, dual MR-DC connectivity is based on radio measurements allowing the master 4G base station MeNB to add or to remove a 5G secondary node SgNB.

For the 5G-SA mode, the mobile is attached to the 5G core network. Even though this book is devoted to the radio access network, the mobile registration, establishment procedures and re-establishment procedures are also explained.

This book describes the architecture of the NG radio access network and the 5G-NR radio interface according to 3GPP (3rd Generation Partnership Project) specifications.

The 5G-NR radio interface was introduced in Release 15 of the 3GPP standards.

The 5G-NR radio interface is the interface between the UE mobile (User Equipment) and the 5GC mobile network (5G Core).

Chapter 1 presents the 5G-NSA and the 5G-SA deployment architecture. It describes the radio access entities and functions of the 5G-SA core network and presents the functions supported by the radio access and the 5GC core network.

In order to prepare RAN virtualization, the radio protocols of the gNB entity are supported by two units: a centralized unit and a distributed unit. This separation creates new interfaces called F1 on the radio access.

Chapter 2 presents the interfaces between the entities of NG-RAN radio access. It describes the protocol architecture on the Xn interface, the NG interface and the F1 interface, as well as the procedures for managing the interfaces of the NG-RAN network.

Similar to 4G CUPS (Control User Plane Separation) evolution, the 5GC network architecture separates the control plane to the user plane. The 5G-NSA and the 5G-SA architectures will be described by detailing interfaces and applications between the radio access entities and between the radio access and the network core (NG interfaces).

Chapter 3 presents general procedures for 5G-NSA and 5G-SA modes. It describes the procedures on the radio access network concerning the cell search, the bearer establishment, the mobility management, the management of the secondary node, the beam management, as well as the procedures to the core network such as attachment and session establishment.

5G-NR spectrum allocation is more flexible compared to 4G LTE. In the time domain, frame flexibility is improved with the introduction of new numerologies. In the frequency domain, flexibility is achieved with bandwidth partitioning allocated to the terminals.

In order to avoid future development constraints, the management of the radio band is refined. Chapter 3 describes the reference signals, physical signals, physical channels and their allocation on the 5G bandwidth.

5G improves the user's throughput with active antenna technologies. Spatial multiplexing improvements are presented in both FR1 and FR2 bands. This chapter describes single- and multi-panel antennas, which require new measurements of the CSI radio channel.

To achieve 20 Gbps in the downlink direction and 10 Gbps in the uplink direction, the 5G-NR radio interface needs the following characteristics:

– the maximum radio channel bandwidth is 100 MHz for the FR1 band and 400 MHz for the FR2 band;

– carrier aggregation allows us to extend the channel bandwidth to 1 GHz;

– 256-QAM (Quadrature Amplitude Modulation);

– MIMO (Multiple Input Multiple Output) 8x8 transmission mode is used in downlink.

Chapter 4 presents the 5G-NR physical layer. It describes uplink and downlink waveform formats, multiplexing modes, the 5G frequency bands and radio frames. Physical signals and physical channels on the uplink and downlink are detailed (uplink physical channels like PRACH, PUSCH, PUCCH and downlink like PBCH, PDSCH, PDCCH). The reference signals are also explained: CSI-RS, DM-RS,

PT-RS, RIM-RS. Chapter 4 also details the mapping of physical channels and physical signals on resource elements.

Chapter 5 presents the physical layer mechanisms to improve throughput and coverage. It describes the SUL mode, the carrier aggregation mechanism, dual connectivity and spectrum sharing for 4G/5G coexistence.

Chapter 6 presents MIMO mechanisms used to improve the robustness of the transmission (diversity), the bit rate (SU-MIMO, MU-MIMO) and the reduction of interference (beamforming). It describes the evolution of antenna capabilities from MIMO to massive-MIMO and single-panel or multi-panel transmission mechanisms. The procedure for managing the beams and improving data rate is described with the use of the code book for the uplink and downlink transmissions.

Chapter 7 presents the partitioning of the 5G band. It describes initial band partitioning and BWP band switching, allowing the mobile to camp on a cell and, under the control of the base station, to adapt its radio capabilities to the traffic load. It also describes the search area and CORESET control resource elements.

Chapter 8 introduces the data link layer. It describes the different SDAP, PDCP, RLC and MAC sublayers by presenting their functional role and the services between the different sublayers. For each of the sublayers, the structure of the protocol is detailed.

Chapter 9 presents the messages sent by the RRC layer: broadcasting information, connection management, measurement configuration. It details the MIB and SIB messages, the procedures for establishing the radio link and the mobility of the mobile, as well as the elements to be measured.

A summary of the content of these nine chapters is provided in the following table.

Chapter	Designation	Content
1	NG-RAN network architecture	Functional architecture NSA/SA Identities UE context gNB-CU/DU QoS and network slicing
2	NG-RAN protocol architecture	Protocol architecture eCPRI functional splitting Interface protocols (Xn, NG, F1) NG-AP and Xn procedure

Chapter	Designation	Content
3	Radio access procedure	Search cell Access procedure Data transmission procedure Session management Secondary node establishment Beam management Attachment
4	Physical layer of the 5G-NR interface	Frequency band Multiplexing structure Reference signals: PSS, SSS, CSI-RS, PT-RS, RIM-RS Physical signals: PSS, SSS Physical channels: PBCH, PDCCH, PDSCH, PMCH, PUSCH, PUCCH, PRACH
5	Operations on frequency bands	Carrier aggregation SUL mode Dual connectivity
6	Multi-antenna structure	SU-MIMO MU-MIMO Beamforming Massive-MIMO Antenna port Channel measurement reports Beam management
7	Bandwidth partition	BWP initial CORESET#0 BWP switching
8	Data link layer	SDAP protocol PDCP protocol RLC protocol MAC protocol Structure protocol
9	RRC protocol	NG-RAN procedure Information systems Connection control Measurements Broadcasting control

April 2021

1

NG-RAN Network – Functional Architecture

1.1. Functional architecture NSA/SA

Unlike previous generations of mobile networks, the deployment of 5G does not require the simultaneous implementation of the 5G core network (5GC) and the NG-RAN (Next-Generation Radio Access Network).

NSA (Non-Standalone Access) and SA (Standalone Access) are two 5G network models:

– SA is a completely new core service-based architecture: each radio node is autonomously controlled by the 5G core network. A service-based architecture delivers services as a set of NFs (Network Functions). NFs in the 5G core network are cloud native;

– NSA relies either on the 4G core network or on the 5G core network. NSA anchors the control signaling to the core network through a radio MN (Master Node). The MN is either a 4G radio node or a 5G radio node. The MN controls an SN (Secondary Node) (4G radio node or 5G radio node) according to the DC (Dual Connectivity) mechanism.

The 5G-SA architecture requires the deployment of a 5G core network connected to the NG-RAN.

The 5G-NSA architecture and the 5G-SA architecture were introduced in Release 15 of the 3GPP standard.

For a color version of all the figures in this chapter, see www.iste.co.uk/launay/5g.zip.

The 5G-NSA configuration implements the MR-DC (*Multi Radio Dual Connectivity*) architecture.

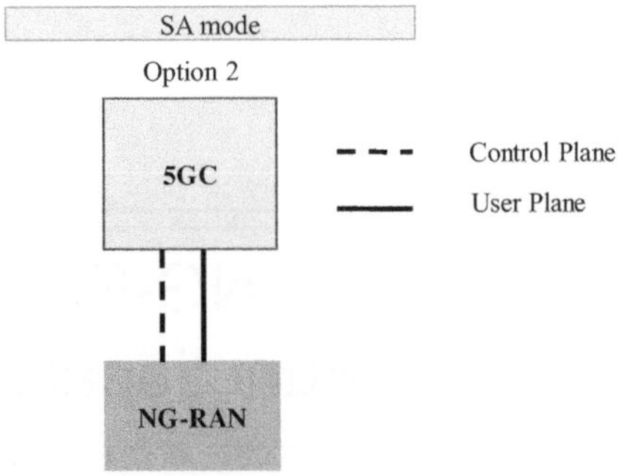

Figure 1.1. *Deployment in the SA mode*

Dual connectivity involves two RAN nodes, i.e. master node (MN) and secondary nodes (SN) which has the following features:

– the MN is connected to the core network for the control plan (signaling) and for the user plane;

– the SN is controlled by the MN. It is connected to the MN for the control plane (C-plane). The user plane (U-plane) is either connected to the MN or connected to the core network;

– the master radio access node controls the secondary radio access node and establishes a bearer, if necessary, for the exchange of data between the two radio nodes.

Dual connectivity defines the “Master Cell Group (MCG) bearer” and the “Secondary Cell Group (SCG) bearer”.

The MCG carries data that will be transmitted on the radio resources allocated by the MN. In the case of carrier aggregation, the MN supports data on the PCell (Primary Cell) and SCells (Secondary Cells).

The SCG carries data that will be transmitted on the radio resources allocated by the SN. In the case of carrier aggregation, the SN supports data on the PCell and SCell.

The split bearer consists of routing the traffic between the MN and the SN. According to the U-plane termination, the split bearer consists of splitting either the MCG bearer or the SCG bearer.

For E-RABs configured as "MCG bearers", the U-plane termination point is located at the MN.

For E-RABs configured as "SCG bearers", the U-plane termination point is located at the SN.

For the core network configuration, each support (MCG, SCG, split bearer) can end on the MN and/or on the SN. The split bearer is transparent for the core network entities.

Several deployment scenarios (Figure 1.2) have been defined for the 5G-NSA:

– option 3: the E-UTRAN access network is connected to the 4G core network. The master node is the 4G radio node (*eNB – evolved Node B*). The secondary node is the 5G radio node (en-gNB). The MR-DC architecture is called *EN-DC* (*E-UTRAN NR Dual Connectivity*);

– option 4: the NG-RAN access network is connected to the 5G core network. The master node is a 5G radio node (*gNB – next generation Node Base Station*). The secondary node is a 4G radio node (ng-eNB). The MR-DC architecture is called *NE-DC* (*NR – E-UTRAN Dual Connectivity*);

– option 7: the NG-RAN access network is connected to the 5G core network. The master node is a 4G radio node (*ng-eNB – next generation eNB*). The secondary node is a 5G radio node. The MR-DC architecture is called *NGEN-DC* (*NG-RAN E-UTRAN NR Dual Connectivity*).

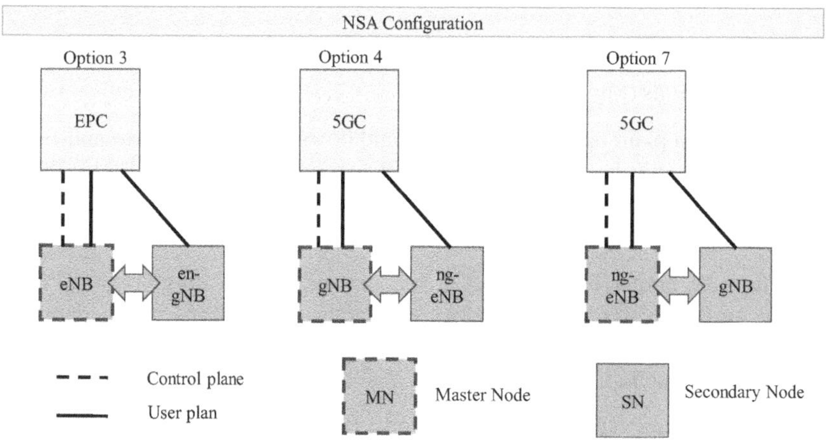

Figure 1.2. *NSA configuration options*

1.1.1. *Option 3*

Option 3 is the non-standalone EN-DC configuration.

Option 3 uses the MN (Master Node) terminated MCG (Master Cell Group) bearer for signaling: the eNB is the master node, and the gNB (gNodeB) acts as the secondary node. The radio access network is connected to EPC.

The 4G base station (eNB) controls the 5G base station (en-gNB) through the X2 interface.

The eNB supports signaling with the MME (*Mobile Management Entity*) through the S1-MME interface and supports the user plane traffic (*MCG bearer*) with the SGW (*Serving Gateway*) entity through the S1-U interface.

The en-gNB base station supports signaling with the eNB. The 5G-NR interface is activated by the eNB over the X2 interface. Once activated, en-gNB controls its own radio resource allocation. The user traffic is either transmitted from the eNB to en-gNB or transmitted from the 4G core network (SGW) over the S1-U interface to en-gNB.

The master node eNB exchanges data in both directions, uplink and downlink, with the mobile.

The secondary node en-gNB allows us to increase both uplink and downlink data rates.

With a DC mechanism, data is transmitted to the mobile according to one of the following variations (Figure 1.3):

– option 3: in plain option 3, all uplink and downlink data flows to and from the LTE part (*MCG split bearer*) of the LTE/NR base station, i.e. to and from the eNB. The eNB then decides which part of the data it wants to forward to the 5G gNB part of the base station over the Xx interface;

– option 3a: both LTE eNG (*MCG bearer*) and 5G-NR en-gNB (*SCG bearer*) exchange traffic to the 4G core network directly. This means that a data bearer allocated to a node cannot share its load over the second node. This option does not suit the case of mobile use;

– option 3x: user data traffic will directly flow to the 5G gNB part of the base station (*SCG split bearer*). The traffic is delivered over the 5G-NR interface to the device, and part of the data can be forwarded over the X2 interface to the 4G eNB.

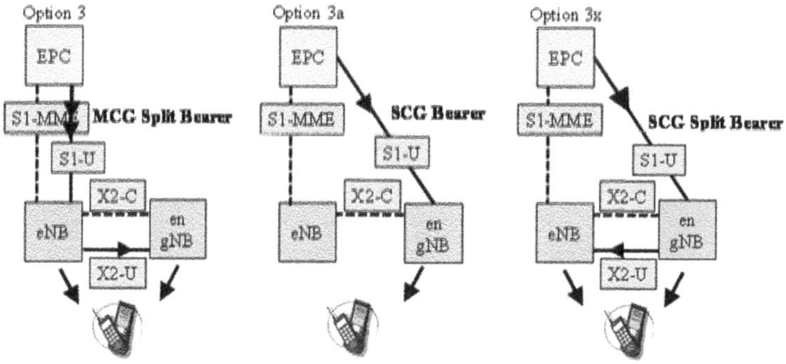

Figure 1.3. *Secondary node addition – option 3*

Option 3 uses the MN terminated MCG bearer for user traffic. The eNB entity splits the S1 bearer into:

– LTE radio support;

– NR support.

Option 3x uses the SN terminated SCG bearer for user traffic. The gNB entity splits the S1 bearer into:

– LTE radio support;

– NR support.

1.1.2. *Option 4*

Option 4 relies on the 5G core (5GC).

The gNB acts as an MN; it supports signaling exchange (*MCG signaling bearer*) with the 5G core network's transport plane through the NG-C interface. The LTE user plane connections go via the 5G-NR through the NG-U interface.

The ng-eNB base station acts as a secondary node. It is controlled by the gNB base station through the Xn-C interface. The ng-eNB is a new generation of the 4G base station.

The gNB controls the ng-eNB through the Xn interface.

The data is transmitted to the ng-eNB entity via one of the following options (Figure 1.4):

– from the master node gNB, which performs the split bearer (option 4, *MN terminated split bearer*);

– from the 5GC network (option 4a, *SCG bearer*).

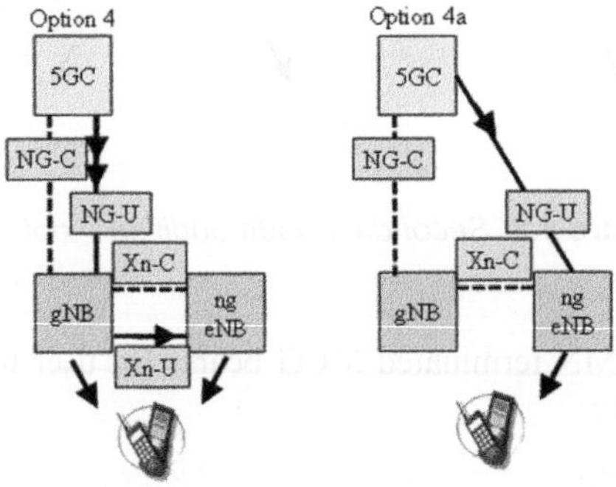

Figure 1.4. *NE-DC architecture – option 4*

1.1.3. *Option 7*

Option 7 relies on the 5G core (5GC).

The ng-eNB acts as an MN; it supports signaling (*MCG signaling bearer*) with the 5GC core network's transport plane through the NG-C interface and exchanges data to the 5G core network's user plane through the NG-U interface.

The gNB base station acts as an SN. It is controlled by the ng-eNB base station via the Xn-C interface.

The ng-eNB (4G base station) controls the gNB through the Xn interface.

The data is transmitted to the gNB entity via one of the following options (Figure 1.5):

– from the ng-eNB base station, which performs a split bearer. This is option 7 (*MN terminated split bearer*);

– from the 5G core network. This is option 7a (*SCG bearer*).

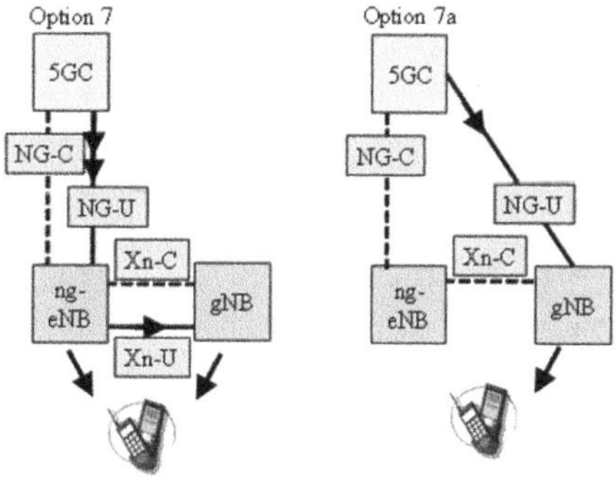

Figure 1.5. *NE-DC architecture – option 7*

1.2. Description of the NG-RAN network

The NG-RAN provides both NR and LTE radio access.

An NG-RAN node is either a gNB (5G base station), providing NR user plane and control plane services, or an ng-eNB (new generation 4G base station) providing the LTE/E-UTRAN services towards the UE (control plane and user plane).

The NG-RAN ensures the connection of mobiles and the reservation of radio resources between:

– the mobile and the ng-eNB base station on a single 4G carrier (LTE) or on several 4G frequency carriers (LTE-Advanced);

– the mobile and the gNB base station on one or more 5G frequency bands (5G-NR).

The gNBs and ng-eNBs are interconnected through the Xn interface. The gNBs and ng-eNBs are also connected, via NG interfaces, to the 5G core (5GC).

The NG interface is the point of reference between the NG-RAN and the 5G core network:

– the NG-C interface is the interface between the radio node and the AMF (Access and Mobility Management Function). It supports signaling via NG-AP (Next Generation Application Protocol);

– the NG-U interface is the interface between the radio node and the UPF (User Plane Function) for tunneling traffic (the IP packet) via GTP-U (GPRS Tunneling Protocol).

The UPF is configured by the SMF (Session Management Function) under the control of the AMF.

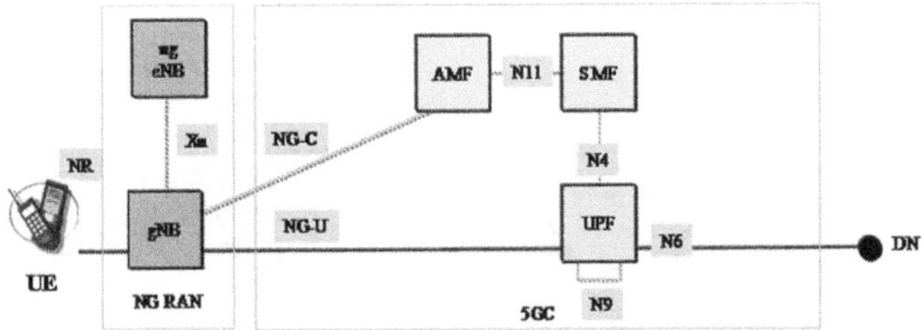

Figure 1.6. *NG-RAN general architecture*

The mobile exchanges data with the DN (Data Network) through logical connections called PDU (Protocol Data Unit) sessions. This logical connection is divided into two parts:

– the NG-RAN ensures the connection of the mobiles with the base station and interconnects the control plane and user plane (traffic) of the mobile UE with the core network;

– the 5G core network interconnects the NG-RAN, provides the interface to the DN, ensures the registration of mobiles, the monitoring of their mobility and the establishment of data sessions with the quality of the corresponding QoS (Quality of Service).

1.2.1. *The NG-RAN*

The NG-RAN provides both an LTE radio interface and a 5G-NR radio interface.

An NG-RAN node is:

– a 5G base station (gNB), which provides the control plane services and the transmission of user plane data through the 5G-NR radio interface;

– an advanced 4G base station (ng-eNB), providing control plane services and data transmission from the user plane to mobiles via the LTE radio interface.

The NG-RAN node is responsible for managing radio resources, controlling the radio bearer establishment of the user plane and managing mobility during the session (handover). The mobile connects to one of the radio nodes.

The NG-RAN node transfers the traffic data from the mobile to the UPF and data from the UPF to the mobile.

When the NG-RAN node receives data from the mobile or from the UPF, it refers to the QFI (QoS Flow Identifier) for the implementation of the data scheduling mechanism.

For outgoing data to the UPF entity, the NG-RAN node performs the marking of the DSCP (DiffServ Code Point) field of the IP (Internet Protocol) header, based on the assigned QFI.

The NG-RAN node performs compression and encryption of traffic data on the radio interface. It can also optionally perform the integrity control of the traffic data exchanged with the mobile.

The NG-RAN node performs the encryption and integrity control of the signaling data exchanged with the mobile on the radio interface.

The NG-RAN node performs the selection of the AMF. The AMF is the function of the core network to which the mobile UE is attached.

The NG-RAN broadcasts the RRC paging received from the AMF.

The NG-RAN node also broadcasts the cell's system information, containing the radio interface characteristics. The devices use these parameters for cell selection and for radio bearer establishment requests.

When a mobile is connected, the NG-RAN uses the measurements made by the mobile to decide on the initiation of a cell change during a session (handover).

In order to manage the services for each connected mobile, the NG-RAN node maintains a UE context information block relating to each mobile. The information saved by the radio node may depend on the mobile usage.

The mobile is either in the RRC connected state (RRC_CONNECTED), the RRC inactive state (RRC_INACTIVE) or the standby RRC state (RRC_IDLE).

When the mobile enters the standby state, the base station is not aware of its presence. Each mobile in the standby state listens to the information broadcasts by the radio node.

There is no UE context at the radio node for the mobile in the RRC_IDLE state.

When the mobile enters the RRC_CONNECTED state or the RRC_INACTIVE state, a mobile radio identifier C-RNTI or I-RNTI, respectively, (Connected/Inactive Radio Network Temporary Identifier) is saved at the radio node. The context of the UE is saved in relation to the RNTI. The context is recorded at the level of the NG-RAN node, which manages the mobile (source node), and it is transmitted to the target node in the event of a handover. The UE context is also created at the level of the MN and at the level of the SN in the event of dual connectivity.

When a mobile is in the connected mode, the NG-RAN node uses measurements made by the mobile to decide whether to trigger a change of node during the session (handover) or to activate or deactivate secondary cells.

1.2.2. *AMF (Access management and Mobility Function)*

The AMF (Access management and Mobility Function) supports:

– the registration of the mobile;

– the access control and the management of mobility on both the NG-RAN and Wi-Fi network access (non-3GPP access);

– network slicing.

The mobile and the AMF exchange data using the NAS (Non-Access Stratum) protocol.

The registration function allows the attachment of the mobile, the detachment of the mobile and the update of its location.

During the attachment, the AMF records the TAI (Tracking Area Identity) location and private identity of the mobile and assigns a 5G-GUTI (5G Globally Unique Temporary Identifier) to the mobile.

5G-GUTI replaces the encrypted private identifier SUCI (Subscription Concealed Identifier) and the private identifier SUPI (Subscription Private Identifier).

Once the attachment procedure is completed, the AMF selects the SMF, according to the DNN (Data Network Name) and the network slice indicator NSSAI (Network Slice Selection Assistance Information).

A load balancing procedure is applied when different SMF can be selected.

The DNN is either communicated by the mobile to the AMF during attachment, or retrieved from the subscriber's profile from the UDR (Unified Data Repository).

The AMF manages a list of TAIs allocated to mobiles, in which the mobile, in the standby state, can move without contacting the AMF to update its location.

The AMF manages the addition and removal of the TNL (Transport Network Layer) association with the entities of the NG-RAN node. In the event of a handover, the source AMF will release the TNL association with the source NG-RAN node and redirect the TNL association to the target NG-RAN node.

1.2.3. *SMF (Session Management Function)*

The SMF (Session Management Function) is responsible for creating, updating and removing PDU (Protocol Data Unit) sessions and managing session context with the UPF (User Plane Function). The SMF injects routing rules to the selected UPFs.

A routing rule corresponds to an entry in the context table of the UPF. This context table contains four fields:

– a correspondence field (PDR (Packet Detection Rule));

– a routing field NH (next hop: IP address, tunnel number TEID (Tunnel End Identifier) or SR (Segment Routing)) to find the next node;

– the quality of service to be applied to the flow (QER (QoS Enhancement Rules));

– the measurement reports to be applied to the flow (URR (Usage Reporting Rules)).

The SMF is responsible for the session management for each DNN and by network slice (S-NSSAI), based on the user profile stored at the UDR.

When requesting a session to be established, the SMF selects a UPF or queries the NRF (Network Repository Function) to obtain the address of the UPF.

The SMF grants an IPv4 or IPv6 address to the mobile. An IP address is provided for each PDU session, based on the address range of the PSA (PDU Session Anchor) selected to join the IP data network. The address range is obtained by either directly querying the selected UPF or by querying the NRF. If the assigned

IPv4 address is a private address, the UPF entity performs NAPT (Network Address and Port Translation) in order to translate the IP address and TCP (Transmission Control Protocol) or UDP (User Datagram Protocol) port numbers.

At the end of the IP session, when the mobile enters the standby state, the SMF releases the session by removing the context at the UPF.

In the event of incoming packets, if the mobile is in the idle state, the SMF sends a notification to the AMF (Downlink Data Notification).

1.2.4. *UPF (User Plane Function)*

The UPF (User Plane Function) manages the routing of user traffic and implements traffic filtering functions.

The PSA UPF is the traffic gateway connecting the 5GC network to the DN (Data Network). The PSA constitutes the anchor point for inter-UPF mobility.

The UPF is the anchor point for traffic when the mobile is moving from one NG-RAN node to another.

The UPF measures the quantity of data consumed for each UE.

The UPF can also implement traffic optimization functions and NAT (Network Address Translation), either from a private IPv4 address to a public IPv4 address, or from an IPv4 address to an IPv6 address and vice versa.

When the UPF receives data from the DN:

– in the absence of a routing context concerning the incoming flow, the UPF informs the SMF. The UPF either stores the data or transmits it to the SMF;

– in the presence of a routing context stored at the UPF level concerning the incoming flow, the flow is either transmitted to an NG-RAN node or to another UPF.

The UPF implements traffic routing rules by configuring the DSCP field of the IP based on the QFI. The QFI is defined by QoS rules which are injected by the SMF to the UPF when establishment of a session is requested. For each incoming piece of data, the UPF performs a traffic inspection (DPI (Deep Packet Inspection)) and classifies the packets into IP flow groups according to the SDF (Service Data Flow) service templates.

The UPF is a branch point supporting the multi-homing function.

The UPF performs replications of the mobile traffic data within the framework of lawful interception.

1.3. Functional separation between the NG-RAN radio interface and the 5G core network

The AMF is in charge of managing the 5G core network and services. It authenticates and registers each mobile and manages their mobility. Once registered, the AMF authorizes services according to the user's profile.

Figure 1.7 summarizes the functions managed on the NG-RAN and on the 5GC.

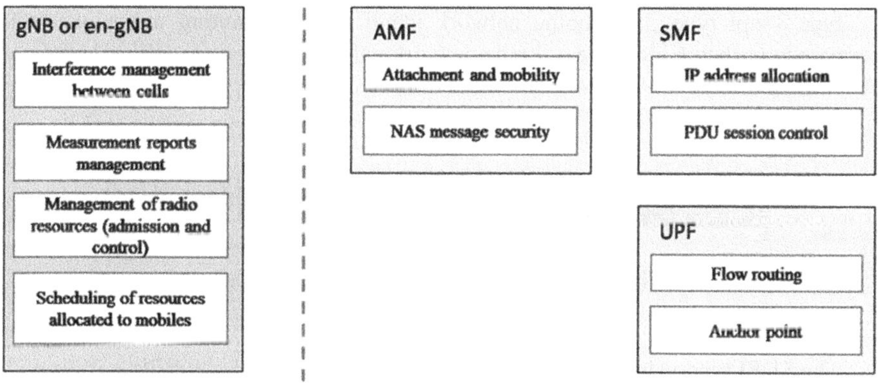

Figure 1.7. *The functional separation between NG-RAN and 5GC*

1.3.1. *Mobile identities*

1.3.1.1. *The identity of the mobile at the level of the AMF*

Registration procedure occurs when the mobile switches on. If authentication succeeds, the state of the mobile changes from the RM-DEREGISTERED state to the RM-REGISTERED state and a user context (UE Context) is created on the AMF.

During the registration procedure, the AMF registers the IMSI (International Mobile Subscriber Identity) from its private identifier SUPI or private and hidden identifier SUCI.

The SUPI and SUCI identifiers allow the core network to identify the subscription associated with the mobile. The identifier format matches with the description of the NAI (Network Access Identifier) in order to be compatible with the DNS (Domain Name Server) servers by respecting the RFC7542 specification.

The SUPI identifier consists of two fields: the type of SUPI identifier (IMSI Identifier (International Mobile Subscriber Identity) or an identifier specific to the operator network) and the IMSI identity value or specific NAI value.

The SUCI identifier is made up of six fields defined as follows (Figure 1.8):

type <supi type>.hni <home network identifier>.rid <routing indicator>.schid <protection scheme id>.hnkey <home network public key id>.out <HPLMN defined scheme out>.

Figure 1.8. *The fields of the SUCI identifier*

The SUPI type value is used to indicate whether the SUPI identifier corresponds to the IMSI identity (type = 0) or a network-specific identifier (type = 1).

The hni field corresponds to the country code (MCC (Mobile Network Country)) and the operator code (MNC (Mobile Network Code)).

The ri field is defined over four digits. The default value is 0, but a specific value is used to identify on which partition of the UDR database the mobile subscription is stored (in the case of UDR, composed of different memory stacks).

The hnkey field identifies the key used for the encryption of the SUCI identifier.

The outfield is the result of the encryption of the mobile's IMSI identity (MSIN (Mobile Subscriber Identification Number)) including the message authentication value (MAC (Message Authentication Code)).

After authentication, the AMF provides a 5G-GUTI. Each 5G-GUTI is unique. This 5G-TMSI identifier is concatenated with the AMF identifier, in order to constitute the 5G-GUTI identifier (Figure 1.9).

The GUAMI identity corresponds to the AMF address. The GUAMI identity is the concatenation of the identity of the PLMN and the identity of the AMF:

<GUAMI> = <MCC><MNC><AMF Identifier>.

The 5G-S-TMSI identifier combines the AMF identity (AMF pool ID and pointer) as well as the 5G-TMSI identifier:

<5G-S-TMSI> = <AMF Set ID><AMF Pointer><5G-TMSI>.

The 5G-S-TMSI identifier is used as a radio identification for paging notifications.

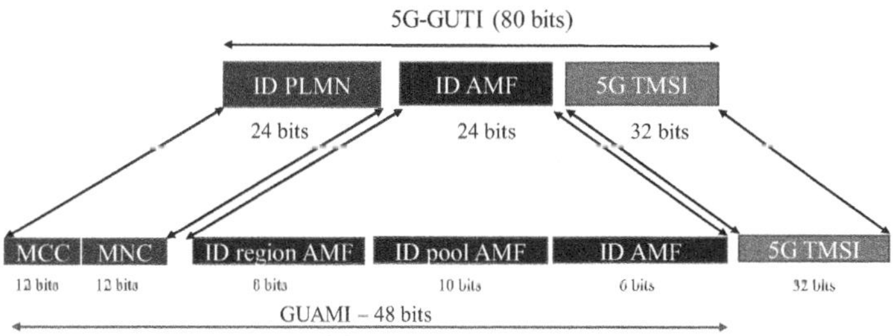

Figure 1.9. *The fields of the 5G-GUTI identifier*

Once registered, the connection state of the mobile at the AMF level is either the connected state (CM-CONNECTED) or the standby state (CM-IDLE).

In the CM-CONNECTED state, a NAS connection is established between the mobile and the AMF function. Because the NAS connection is encapsulated in the RRC message, the mobile is both in the CM-CONNECTED state with the AMF and in the RRC_CONNECTED state with the NG-RAN node. The NG-RAN node creates a UE context with RNTI (Radio Network Temporary Identifier).

After a period of inactivity, the NG-RAN node can:

– suspend the radio connection. The mobile goes to the RRC_INACTIVE state, the context is kept by the NG-RAN node and the UE context is kept at the AMF level (the mobile is still in the CM-CONNECTED state);

– release the radio bearer and remove the UE context. The mobile goes to the RRC_IDLE state and the UE context goes to the CM_IDLE state at the AMF level.

In the RRC_IDLE state, the mobile listens to the information sent by the NG-RAN node. In the RRC_IDLE state, cell reselection is managed by the mobile.

1.3.1.2. *The identity of the mobile at the level of NG-RAN*

Each NG-RAN node manages a set of mobiles. The NG-RAN node assigns a specific RNTI radio identifier to each mobile. A broadcast identifier is also used to broadcast information like the common control channel and information system.

Consequently, for any mobile in the standby mode, the NG-RAN node uses the following radio identities:

– P-RNTI (Paging) to send a paging request;

– SI-RNTI (System Information) to broadcast SIB messages;

– RA-RNTI (Random) to identify a mobile when requesting radio access (random access procedure; see Chapter 3);

– TC-RNTI (Temporary Cell) which helps to retrieve the information exchanged during the connection procedure, following the random procedure.

If the mobile is in the RRC_INACTIVE state (in the standby mode) on the NG-RAN, its identifier is called I-RNTI.

If the mobile is in the RRC_CONNECTED state, connected with the NG-RAN node, then several radio identities are used per mobile:

– C-RNTI (Cell): unique identification of the mobile for the RRC connection and for scheduling;

– CS-RNTI (Configured Scheduling): radio identifier of the mobile for the RRC connection and for semi-persistent scheduling;

– SP-CSI-RNTI (Semi-Persistent): radio identifier of the mobile used for the transmission of radio channel information (CSI: Channel State Information) on the uplink traffic channel with semi-persistent scheduling;

– MCS-RNTI (Modulation Coding Scheme): radio identifier of the mobile to indicate the modulation and coding scheme (MCS) of the mobile for the transmission of data on the downlink and uplink traffic links;

– TPC-RNTI (Transmit Power Control RNTI) is used to retrieve the information corresponding to the power control for both the traffic and uplink control channels.

1.3.2. *Mobile mobility*

1.3.2.1. *Mobility in the idle mode*

When the mobile is in the idle state, it is located by the AMF from the TAI (Tracking Area Identity) value.

The mobile listens to the broadcast information and to the paging information sent by the NG-RAN node.

If the mobile is in the discontinuous reception mode (DRX), the frame that carries the paging notification is calculated from the 5G-identity S-TMSI of the UE modulo 1024.

The mobile becomes aware of the paging when the content of the paging message carries the 5G-S-TMSI identifier or the I-RNTI identifier of the mobile.

The mobile informs the AMF of a change of tracking zone when it falls under the coverage of a radio node whose TAI is different from the previous TAI coverage zone.

The selection or reselection of cells is based on the transmitted synchronization signal CD-SSB (Cell Defining Synchronization Signal and PBCH Block) specific to each cell.

1.3.2.2. *Mobility in RRC_Inactive*

The mobile is in the RRC_INACTIVE state for the NG-RAN and in the CM_CONNECTED state for the 5G core network. The mobile context is always stored on the node where the mobile was connected.

The mobile can move in the RAN Notification Area (RNA) without notifying the NG-RAN nodes. An RNA zone comprises one or several cells located in the same TAI zone.

In the case of mobile terminating data, the UPF transfers the packet to the last NG-RAN node on which the mobile had established a radio connection. The NG-RAN node broadcasts a paging request over the air interface and transmits (through the Xn interface) a paging order to neighboring NG-RAN nodes that are configured on the same RNA area. If the paging request fails (the mobile is not reachable), then the NG-RAN node informs the AMF.

If the mobile establishes a radio connection with an NG-RAN node other than the last serving node, then the new NG-RAN node initiates a procedure to retrieve the context of the mobile UE from the old serving node. In the case where the request fails, the NG-RAN node triggers a new RRC connection.

1.3.2.3. *Mobility in the RRC_Connected state*

When the mobile is in the RRC_CONNECTED state and CM_CONNECTED state, its mobility is controlled by the NG-RAN node via the handover mechanism between a source radio node and a target radio node, or between two beams of the same radio node.

In the case of a handover, the NG-RAN node exchanges RRC signaling with the mobile. There are several types of handover:

– on the Xn interface between two neighboring connected nodes;

– on the NG interface with the UPF.

In the case of beam-based coverage, the beam selection is carried out at the MAC (Medium Access Control) layer of the mobile from mobile measurements. No RRC signaling occurs when the beam is changing since the mobile was configured at the beginning of the radio connection in terms of the measurements to be carried out (Measurement Objects).

The mobile makes radio measurements on one or more beams of a cell and determines:

– the radio quality of the beam by filtering the measurements on the L1 layer;

– the quality of the radio link of the cell by averaging the measurement of the different beams at the level of the L3-RRC layer.

The mobile also carries out measurements on the quality of the SSB block (Synchronization Signal and Broadcast) of the beams under the coverage of the cell, and also the quality of the neighboring intra- and inter-band cells. Each SSB measurement is seen by the mobile as a different cell.

Table 1.1 summarizes the different cases.

<table>
<tr><td>RRC_IDLE</td><td>RRC_INACTIVE</td><td>RRC_CONNECTED</td></tr>
<tr><td colspan="2">Cell selection is controlled by the UE in the function of radio access network parameters (AS: Access Stratum).</td><td rowspan="4">UE mobility is known by the core network.
The core network knows the identity of the radio node with which the terminal is connected.</td></tr>
<tr><td colspan="2">The UE listens to broadcast signal.</td></tr>
<tr><td>Paging notifications are initiated by the core network.</td><td>Paging notifications are initiated by the NG-RAN node.</td></tr>
<tr><td rowspan="2">The mobile has a temporary identity created by the core network.
No UE context is stored on the NG-RAN node.</td><td>The NG-RAN node knows the RNA location area on which the mobile is camped</td></tr>
<tr><td colspan="2">The mobile radio context is stored at the mobile side and the NG-RAN node side. The core network and the NG-RAN node exchange information through NG-C and NG-U interfaces.</td></tr>
</table>

Table 1.1. *RRC mobile states*

1.4. Scheduling and QoS

1.4.1. *Scheduling*

Scheduling enables us to share radio resources among all connected users. A scheduling task is performed in real time to share radio resources among all mobiles. Scheduling attribution is calculated based on:

– the quality of the radio link for each mobile;

– requirements in terms of the quality of service expected by each mobile;

– the state of the mobile buffer.

The scheduler is performed at the MAC layer of the NG-RAN node and defines the scheduling for downlink and uplink transmissions.

The quality of the radio link is used to define the modulation and coding scheme (MCS) for a given mobile, as well as the transmission power. The MCS is dynamically adjusted according to the HARQ (Hybrid Automatic Repeat Request) retransmission rate.

To meet mobile requirements, several strategies can be defined at the scheduler level:

– fairness, a strategy for which each mobile receives the same resource allocation regardless of the modulation scheme;

– proportional fairness which allocates more frequency resources for mobiles with less efficient MCS;

– round-robin, a strategy which consists of allocating equal resources to all mobiles;

– max-CQI, a strategy that aims to maximize cell capacity by prioritizing the allocation of radio resources to mobiles which have the best radio quality.

The scheduler is based on measurement reports:

– the state of the buffer BSR (Buffer Status Report);

– the quality of the radio link (CSI-RS);

– the rising power margin PHR (Power Headroom Reports);

– Inter-Cell Interference Coordination (ICIC) between NG-RAN nodes.

From this information, the scheduler defined:

– the frequency radio resources to be allocated for each mobile;

– the number of spatial layers that can be used, depending on the category of the mobile;

– the transmission TTI (Time Transmission Interval) instants.

The mobile listens to the information transmitted over the PDCCH (Physical Downlink Control Channel) logical control channel and decodes the information channel when it detects its radio identifier C-RNTI.

The scheduler makes its decisions at each slot. The duration of the TTI slot depends on the spacing between SCS (SubCarrier Spacing). While the scheduler decision is 1 ms in 4G, it is variable from 1 ms, 500 μs or 250 μs for 5G on the FR1 (Frequency Range 1) band, and can go down to 125 μs or 62.5 μs for the FR2 band.

Transmission on the downlink direction can be pre-empted for critical communication (low latency). The NG-RAN node informs the mobiles by transmitting the INT-RNTI identifier over the PDCCH control channel.

Semi-persistent scheduling allows us to periodically allocate radio resources for a mobile. The periodicity of the messages is transmitted over the RRC layer and the resource allocation is transmitted to the mobile via the CS-RNTI identifier.

1.4.2. *Support for quality of service on radio link*

The QoS (Quality Of Service) control consists of implementing the maximum quality of service applicable to a data flow.

Like the 4G mobile network, only the core network is aware of the service requirements: QoS management is under the control of the core network (AMF). The NG-RAN has no knowledge of the service to be managed. Thus, when establishing a PDU session, the AMF entity establishes QoS rules between the radio node and one or more UPF entities.

The PDU session carries IP flows with one or more different qualities of service for all flows. Each flow is associated with a QFI flow profile identifier. The flow profile corresponds to a QoS indicator (5QI: 5G QoS Identifier) and an allocation and retention priority (ARP). The QFI flag is unique within a PDU session. The flag is either configured during the PDU session establishment procedure or during the PDU session modification procedure.

The value of a QFI is configured by the AMF during the procedure of session establishment; the AMF querying the unified UDR database to know the user's authorized QoS. For the establishment of dedicated services, the SMF chooses the QoS characteristics (5QI/ARP) according to the values stored at the SMF or by querying the PCF entity. The 5QI/ARP combination defined by the PCF link in a PDU session is a QoS flow binding.

The 5QI indicator is a parameter standardized by the 3GPP standard, allowing us to define:

– the type of resource (Guaranteed Bit Rate or not): GBR or non-GBR;

– priority;

– the maximum transmission delay within the 5GS mobile network;

– the residual error rate.

The standardization of the 5QI value makes it possible to indicate how the flow is processed on each element of the user's plan, in order that processing is consistent between the entities of the 5G core network and of the NG-RAN access.

The 5QI indicator is identical to the QCI (QoS Class Identifier) indicator for the 4G network for non-critical services (indicator from 1 to 80). New QCI values (81–85) have been defined in the case of URLLC services to guarantee speed and critical delay (delay critical GBR).

5QI	Type	Priority	Packet Delay Budget	PLER	Services
1	GBR	2	100 ms	10^{-2}	Conversational Voice
2		4	150 ms	10^{-3}	Conversational Video (Live Streaming)
3		3	50 ms	10^{-3}	Real Time Gaming, V2X messages
4		5	300 ms	10^{-6}	Non-Conversational Video (Buffered Streaming)
65		0.7	75 ms	10^{-2}	Mission Critical user plane Push To Talk voice (e.g. MC-PTT)
66		2	100 ms	10^{-2}	Non-Mission-Critical user plane Push To Talk voice
75		2.5	50 ms	10^{-2}	V2X messages
5	Non-GBR	1	100 ms	10^{-6}	IMS Signaling
6		6	300 ms	10^{-6}	Video (Buffered Streaming) TCP-Based (e.g. www, email, chat, ftp, p2p, etc.)
7		7	100 ms	10^{-3}	Voice, Video (Live Streaming), Interactive Gaming
8		8	300 ms	10^{-6}	Video (Buffered Streaming) TCP-Based (e.g. www, email, chat, ftp, p2p, etc.)
9		9	300 ms	10^{-6}	Video (Buffered Streaming) TCP-Based (e.g. www, email, chat, ftp, p2p, etc.). Typically used as default bearer
69		0.5	60 ms	10^{-6}	Mission Critical delay sensitive signaling (e.g. MC-PTT signaling)
70		5.5	200 ms	10^{-6}	Mission Critical Data (e.g. example services are the same as QCI 6/8/9)
79		6.5	50 ms	10^{-2}	V2X messages
80		6.8	10 ms	10^{-6}	Low latency eMBB applications (TCP/UDP-based); Augmented Reality
82	Delay Critical GBR	1.9	10 ms	10^{-4}	Discrete Automation (small packets)
83		2.2	10 ms	10^{-4}	Discrete Automation (big packets)
84		2.4	30 ms	10^{-5}	Intelligent Transport Systems
85		2.1	5 ms	10^{-5}	Electricity Distribution-high voltage

Table 1.2. *5G QoS characteristics*

The ARP parameter allows the NG-RAN node to choose whether the bearer establishment request should be made or rejected in the event of congestion.

The QFI value is coded on 6 bits. The 5QI value is set between 1 and 85. For any 5QI value less than or equal to 64, the QFI indicator and 5QI can be the same.

When the mobile is in the RRC_CONNECTED state, the management of QoS rules is delegated to the 5G-NR radio interface.

A user's plane traffic in a PDU session with the same QFI flag is handled with the same traffic routing rules (e.g. sequencing rules, admission level).

The role of the radio node is to configure one or more radio data bearers (RAB: Radio Access Bearer) and to perform a mapping between the QFI and the bearer(s) from a TFT flow filtering template (Traffic Flow Template).

For uplink, there are two ways to control the mapping between the radio bearers and the QoS of IP flows:

– reflective QoS for which the mobile replicates QoS rules received in downlink (configuration of the TFT flow policy rules);

– explicit configuration for which the uplink QoS configuration is defined by configuring the radio bearer.

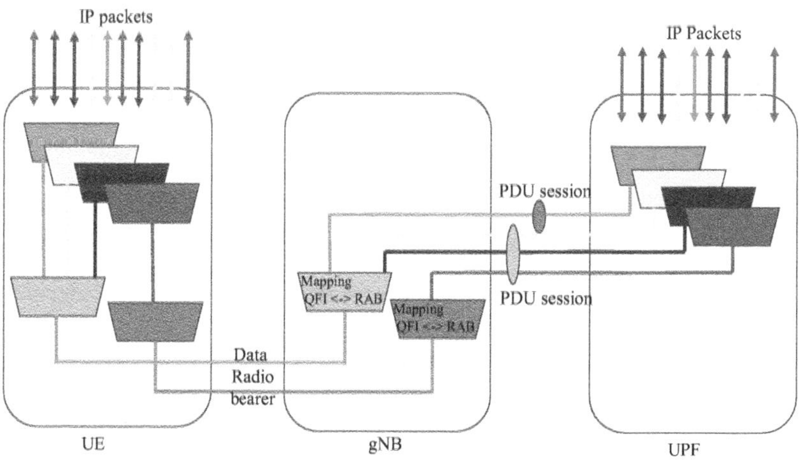

Figure 1.10. *QFI management in the user's plane*

1.5. Security architecture

The security architecture implemented on the 5G mobile is based on:

– mutual authentication between the 5GC core network and mobile (UICC);

– ciphering and integrity of NAS signaling messages exchanged between the mobile and the AMF;

– AS security through the 5G-NR radio interface between the mobile and the NG-RAN node. Security concerns the integrity control and encryption of RRC messages and IP packets. Integrity on IP packets is optional.

Data integrity:

– ensures that the data have not been altered by a third party between transmission and reception;

– verifies the transmitting source;

– ensures that a message already received is not reused.

Encryption ensures the confidentiality of data exchanged between two entities.

The security of the NAS and AS messages consists of deriving different keys at the level of the mobile and at the level of the following entities (Figure 1.11):

– The AMF:

– K_{AMF} key;

– K_{NASint} key from the K_{AMF} key for the integrity check of the NAS signaling;

– K_{NASenc} key from the K_{AMF} key for the encryption of the NAS signaling.

– The radio node:

– K_{gNB} key from the K_{AMF} key;

– K_{RRCenc} key derived from the K_{gNB} key for the encryption of RRC signaling on the 5G-NR interface;

– K_{RRCint} key derived from the K_{gNB} key for the integrity check of RRC signaling on the 5G-NR interface;

– K_{UPenc} key derived from the K_{gNB} key for encrypting IP traffic on the 5G-NR interface;

– optionally, a K_{UPint} key derived from the K_{gNB} key for the integrity check of IP traffic on the 5G-NR interface.

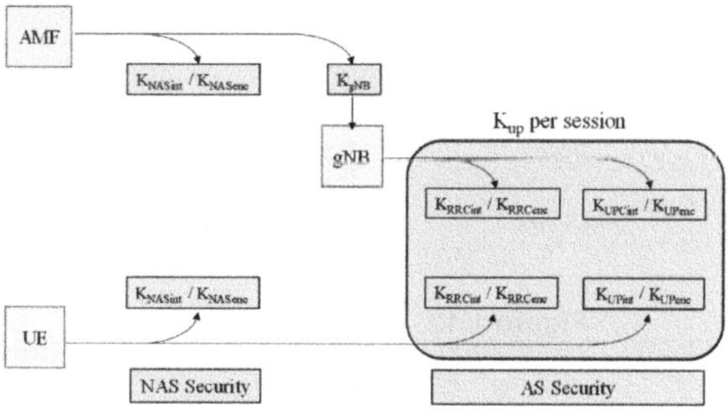

Figure 1.11. *Security architecture*

The mobile must support the NAS security based on information transmitted by the 5G core network and AS security, according to the indications sent by the NG-RAN access node.

5G security is based on the use of:

– NEA encryption algorithms (Encryption Algorithm for 5G);

– NIA (Integrity Algorithm for 5G) integrity control algorithms;

– the K_{UPenc}, K_{RRCenc}, K_{NASenc} encryption keys consist of 128 bits.

The encryption and integrity control algorithms are similar to those used on the LTE interface:

– NEA0/NIA0: no ciphering;

– 128-NEA1/128-NIA1: algorithm SNOW 3G (flow ciphering);

– 128-NEA2/128-NIA2: algorithm AES (bloc ciphering);

– 128-NEA3/128-NIA3: algorithm ZUC (flow ciphering).

Encryption and integrity are based on the following parameters:

– a 32-bit counter;

– the identity of the 5-bit bearer;

– the direction of the connection on one bit;

– the length of the message.

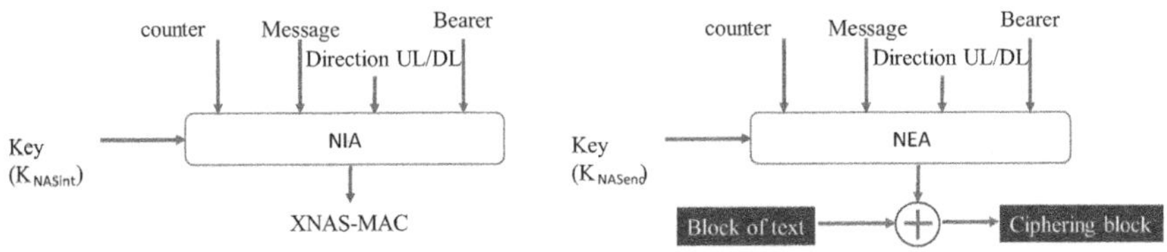

Figure 1.12. *Ciphering and integrity*

1.6. Network slicing

Network slicing is the embodiment of the concept of running multiple logical networks as virtually independent business operations on a common physical infrastructure in an efficient and economical way.

Virtualization is a hardware abstraction to partition network resources into distinct logical segments.

Network partitioning makes it possible to allocate a part of server hardware resources (NFVI: Network Function Virtualization Infrastructure) to network functions (VNF: Virtualized Network Functions).

Hardware capabilities are dynamically managed based on the number of users, on the one hand, and the profile of each user, on the other hand. By default, the 3GPP standard has defined four types of services:

– eMBB: evolved Mobile BroadBand to manage smartphone services such as high speeds, several session establishments, handover management, low latency;

– mMTC: massive Machine Type Communication to manage the sessions of IoT terminals (low speed, little transmission and mainly in the upstream direction, long delay);

– URLLC: Ultra-Reliable Low-Latency communication for critical communications requiring very low latency (less than 1 ms for the user plane) and efficient management of the handover;

– V2X: Vehicle to Everything for autonomous vehicles (between vehicles, with radio infrastructure, etc.).

Network slicing provides all the functionality of the 5G network, including optimization of the radio access network and core network entities to meet the service level agreement (SLA) requirements requested by the user.

Virtualization allows us:

– to allocate a set of material resources (storage capacity, network performance, computing capacity in terms of the number of CPUs);

– to deploy optimized software instances on the hardware resources. The instances correspond to the NFV network functions to be deployed:

- in the radio access network by dividing the radio functions into two entities gNB-CU and gNB-DU,

- in the 5G core network (AMF, SMF, PCF, etc.),

- to deploy optimized network functions (content cache, video optimizer, malware detection, etc.).

The set of hardware and software resources form a Network Slice Instance (NSI). The network instance is split into an RAN Slice Instance (RSI) and a Core Network instance.

From a user point of view, the mobile requests registration on a network instance from the 5GS network. The mobile profile allows the network to define the optimized network instances through the S-NSSAI (Single Network Slice Selection Assistance Information) identifier.

The S-NSSAI indicator is composed of two fields:

– SST: Slice Service Type defined user profile (1: eMBB, 2: URLLC, 3: mMTC, 4: V2X);

– SD: Slice Differentiator to differentiate specific services within an SST service type.

The S-NSSAI indicator is stored at the UDM database for each user profile and stored in the mobile. Each mobile can subscribe to up to eight S-NSSAI. S-NSSAI indicators are integrated into the NSSAI indicator.

When requesting registration, the mobile sends the desired NSSAI flag in the RRC request. The gNB-CU entity selects the AMF entity from the NSSAI indicator if possible; otherwise, it selects a default AMF entity. The AMF entity consults the UDM entity to know the value of the NSSAI indicator that will actually be implemented. The answer depends on the customer's profile and the radio access network (3GPP, non-3GPP or roaming).

At the 5GC core network, when the mobile wants to establish a logical connection for a data service, it sends the RRC Service Request to the AMF with the S-NSSAI flag. The AMF selects the most suitable SMF.

On NG-RAN, the gNB-CU entity selects the gNB-DU entity based on the S-NSSAI flag. The division of the network on radio access is defined by:

– common RRC functions (management of sharing of radio resources between slices) and specific RRC functions (DRX, eDRX, timers, QoS, etc.);

– PDCP, RLC, common or specific functions (header compressions, acknowledgment, etc.);

– sequencing (MAC function) and prioritization on the physical layer.

The objective of virtualizing the network is to provide network flexibility and dynamic adaptation to the needs of different users, in order to meet performance indicators specific to the services requested (latency, throughput, packet loss, etc.). This flexibility is provided by an orchestrator which supervises the network functions and delegates the traffic management to the radio controllers (SD-RAN – Software Defined RAN) and network controllers (SDN: Software Defined Networking). In addition, virtualization allows isolation of the different slices.

1.7. References

All standards can be downloaded from the ETSI website: https://www.etsi.org/standards.

3GPP TS 37.340
Universal Mobile Telecommunications System (UMTS); LTE; 5G; NR; Multi-connectivity; Overall description; Stage-2
Version 15.3.0 Release 15

3GPP TS 38.401
5G - NG-RAN: Architecture description
Version 15.2.0 Release 15

3GPP TS 24.501
5G - Non-Access-Stratum (NAS) protocol for 5G System (5GS); Stage 3
Version 15.0.0 Release 15

3GPP TS 23.003
Digital cellular telecommunications system (Phase 2+) (GSM); Universal Mobile Telecommunications System (UMTS); Numbering, addressing and identification
Version 15.5.0 Release 15

3GPP TS 33.501
5G - Security architecture and procedures for 5G System
Version 15.2.0 Release 15

2

NG-RAN Network – Protocol Architecture

2.1. The protocol architecture of the radio interface

The NG-RAN radio access network is divided into two strata (Figure 2.1):

– the access stratum AS includes air interface and provides functions related the to data link layer and the lower part of OSI layer 3. The access stratum deals with the radio protocols on the Uu interface and with the protocols managed on the NG (Next Generation) terrestrial interface;

– the non-access stratum (NAS) deals with the exchange of traffic and data between the mobile and the core network. As a result, NAS is concerned with all the protocols allowing the exchange of signaling between the mobile UE (User Equipment) and the AMF (Access and Mobility Management Function) entity of the 5G core network. The NAS protocol refers to the procedure related to mobility, authentication and management of service requests (data session or Voice over IP call).

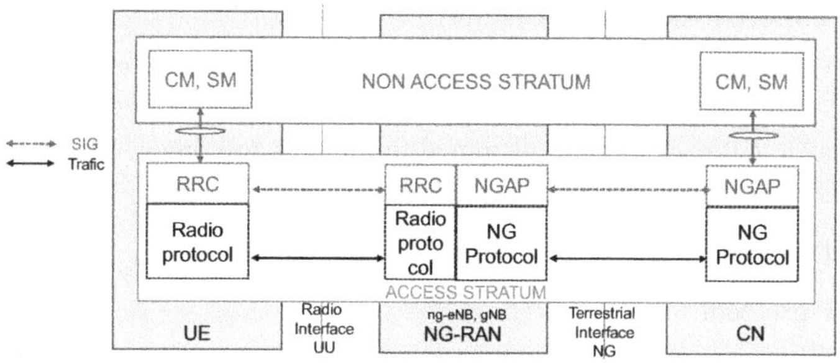

Figure 2.1. *Protocols on the 5G interface*

For a color version of all the figures in this chapter, see www.iste.co.uk/launay/5g.zip.

The protocols on the Uu and NG interface concern:

– the user plane protocols used to transport the traffic data of the PDU (Protocol Data Unit) session;

– the control plane protocols allowing us to control the PDU sessions, the radio connection and the mobility during a session (handover).

2.1.1. *Protocol stack on the Uu interface*

The Uu radio interface is the interface between the mobile and the NG-RAN access node.

The NG-RAN access network includes two types of radio entities:

– the ng-eNB (next generation evolved Node Base station) radio station offering 4G radio access;

– the gNB radio station (next generation Node Base station) offering 5G radio access.

The Uu interface supports:

– the RRC (Radio Resource Control) signaling exchanged between the mobile and the NG-RAN node;

– the user traffic transmitted in the DRB (Data Radio Bearer).

Signaling data (RRC messages) is transmitted to the data link layer. The data link layer is divided into three sublayers:

– PDCP protocol (Packet Data Convergence Protocol);

– RLC protocol (Radio Link Control);

– MAC (Medium Access Control) protocol.

User traffic (IP packets) is transmitted to the data link layer. The data link layer is divided into four sublayers (Figure 2.2):

– SDAP protocol (Service Data Application Protocol);

– PDCP protocol;

– RLC protocol;

– MAC protocol.

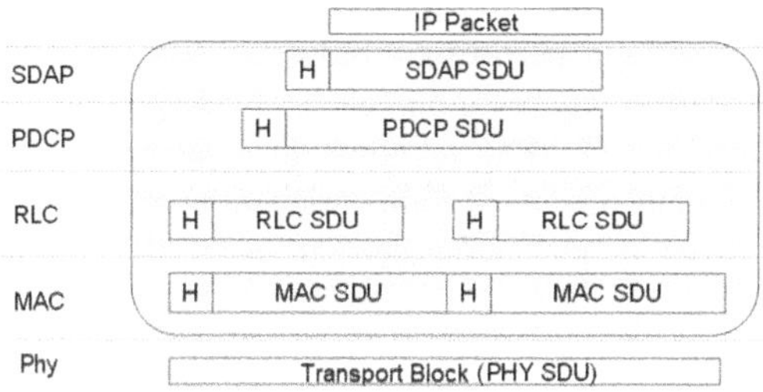

Figure 2.2. *Processing of IP packet in the DataLink layer*

The SDAP protocol provides a mapping between the quality of service (QoS) associated with the IP flows and the radio media transporting this flow between the mobile and the NG-RAN node.

The PDCP sublayer performs IP header compression and decompression through the use of the RoHC (Robust Header Compression) mechanism. The PDCP sublayer implements the security functions by encryption/decryption, as well as protection of the integrity of the data received (IP packets and signaling). Finally, the PDCP sublayer manages the duplication of the transmitted PDCP frames, the reorganization of the frames (in particular for services of the URLLC – Ultra-Reliable Low-Latency Communication types), the reorganization of the packets and the detection of duplicated frames.

In the case of the 5G-NSA option 3X (dual connectivity), the PDCP sublayer of the 5G base station separates the incoming traffic bearer from the core network to the RLC sublayers of the two base stations (split-bearer between the MN: Master Node eNB and SN: Secondary Node en-gNB).

The RLC sublayer provides data segmentation services for the PDCP sublayer. There is one RLC channel for each radio bearer configured for the mobile. The RLC sublayer provides three modes of data transmission: acknowledged mode (AM), transparent mode (TM) and unacknowledged mode (UM). In the case of the acknowledged mode, the errors are corrected via the ARQ (Automatic Repeat reQuest) mechanism.

In order to reduce latency compared to LTE-RLC sublayer, the NR-RLC sublayer no longer performs:

– data concatenation;

– data reorganization, this function is delegated to the PDCP sublayer.

The MAC sublayer multiplexes the logical channels to optimize the physical resources of the radio layer and performs rapid error correction (H-ARQ: Hybrid ARQ). In the case of carrier aggregation, each H-ARQ entity is independently implemented and there is one H-ARQ acknowledged per cell. The MAC sublayer also manages the sequencing of data by allocating radio resources to the mobiles according to the QoS parameters requested at each TTI (Time Transmission Interval).

The MAC sublayer manages the multiplexing of logical channels into transport channels which are then multiplexed into physical channels.

Three types of channels are defined (Figure 2.3):

– the logical channel defines the structure of the data at the interface between the RLC and MAC sublayers;

– the transport channel defines the structure of the data at the interface between the MAC sublayer and the physical layer;

– the physical channel defines the data structure between the two parts constituting the physical layer: on the one hand, coding, and, on the other hand, modulation and multiplexing.

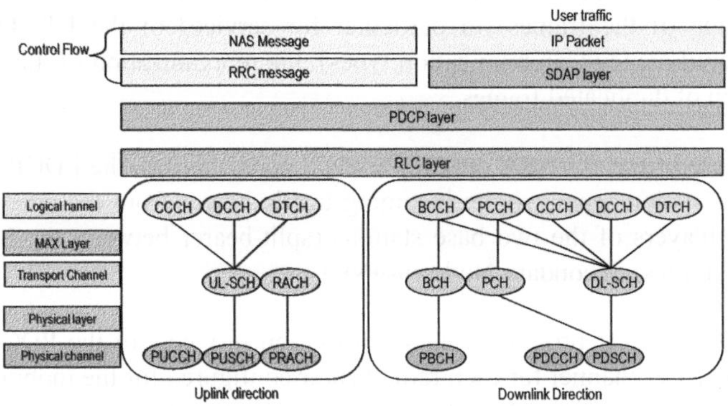

Figure 2.3. *The structure of the radio interface*

The physical layer (PHY) manages the encoding/decoding of data (user plane and control plane), OFDM modulation/demodulation, and the distribution of flows by antenna with digital precoding to allow different transmission modes (SISO, SIMO, MISO, MIMO). The RF radio chain transposes the baseband signal to an RF signal and vice versa.

The 5G physical layer is described by the NR (New Radio) access technology.

2.1.2. *The protocol architecture on the Xn interface*

An NG-RAN node is either a gNB (i.e. a 5G base station) providing NR control plane services and user plane traffic or an ng-eNB providing LTE services towards UE.

Like 4G base stations, an NG-RAN node carries resource management and logic control functions (i.e. Radio Resource Management, radio admission control, etc.) and can communicate directly with another node.

NG-RAN nodes are interconnected via the Xn interface in order to:

– manage user mobility between NG-RAN nodes;

– increase the transmission rate by the dual connectivity mechanism;

– manage interference management (ICIC: Inter Cell Interference Coordination);

– allow optimization of radio settings (SON: Self Optimization Network).

The Xn interface is the reference point between two entities of the NG-RAN node which belong to the same area of tracking. The Xn interface is not mandatory. Signaling is exchanged through the XnAP protocol and user traffic plane through the GTP-U (GPRS Tunnel Protocol User) protocol.

The protocol stack of the Xn interface is similar to the protocol stack of the X2 interface. It supports the exchange of traffic and the exchange of signaling between two nodes (Figure 2.4).

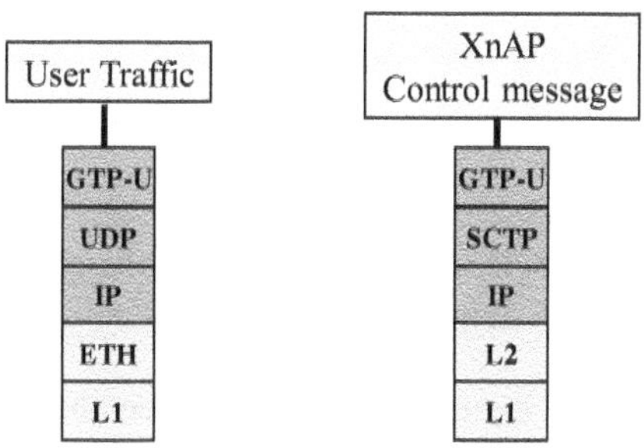

Figure 2.4. *The protocol stack of the Xn interface*

The Transport Network Layer (TNL) uses:

– GTP-U and UDP (User Datagram Protocol) over IP (Internet Protocol) for data transmission;

– the SCTP protocol (Stream Control Transmission Protocol) over IP for the control information.

The SCTP protocol provides reliable data transport by retransmitting erroneous data. It detects data loss and data duplication.

The tunnel established between two NG-RAN nodes is unidirectional and is temporarily established for the transfer of data during a handover or for the dual connectivity mechanism.

An NG-RAN node can be split (Figure 2.5) into an NG-RAN CU (Centralized Unit) and one or more NG-RAN DU (Distributed Unit).

The Xn-C interface is the logical interface between the gNB entity and the gNB-CU.

The E1 interface is the logical interface between two gNB-CUs.

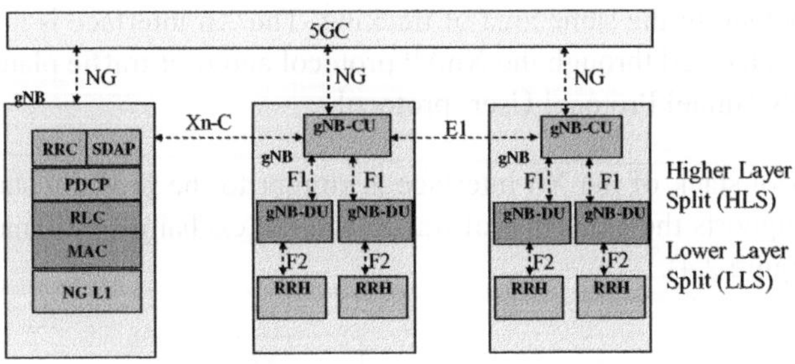

Figure 2.5. *Interfaces between gNB entities and gNB-CU/gNB-DUs*

The F1 interface is the logical interface between the gNB-CU and the gNB-DU. It supports signaling via F1-C functions and traffic via F1-U functions.

The F2 interface is the logical interface between the gNB-DU and the RRH (Remote Radio Head). The data is transported by a serial bus respecting the eCPRI (evolved Common Public Radio Interface) protocol on the optical fiber, also called RoF (Radio Over Fiber).

2.1.2.1. *Control plane functions*

The XnAP control plane on the Xn-C interface supports:

– functions used for management of signaling associations between NG-RAN nodes, surveying the Xn interface and recovering from errors (i.e. reset function after abnormal failure);

– functions to allow initial setup, modification or release of the Xn interface between two NG-RAN nodes;

– functions to allow two NG-RAN nodes to update application level data at any time, including activation or deactivation of the interface within the framework of an energy management procedure;

– function to allow exchange of information between the source and target NG-RAN nodes to initiate a handover or function informing an existing target NG-RAN node that a handover will ultimately not take place;

– function to exchange the context of a mobile between two NG-RAN nodes;

– function to manage paging procedure when the mobile is in the RRC_INACTIVE state;

– functions to establish or release a tunnel and data transfer between two NG-RAN nodes.

2.1.2.2. *User plane functions*

User plane functions (UPF) on the Xn-U interface support:

– data transfer in the case of dual connectivity or the handover mechanism;

– user plane flow control allowing an NG-RAN node receiving data from a second NG-RAN node to provide information associated with the transfer of the flow;

– fast retransmission in the event of failure of one of the nodes. The messages of the XnAP protocol ensure coordination between the node hosting the PDCP sublayer and the corresponding node, to allow the transmission of data under good RF radio conditions by excluding the faulty node.

2.1.3. *Protocol architecture on the F1 interface*

Several options define the functional split between the gNB-CU and gNB-DU modules. The two main criteria that determine the choice of the option are interface throughput and latency.

The 3GPP standardization proposes a division of the functionalities of the gNB entity on the NG-RAN access network and of the ng-eNB entity on the E-UTRAN access network into two units: a distributed unit (DU: Distributed Unit) and a centralized unit (CU: Centralized Unit).

In Release R.15, RRC, SDAP and PDCP protocol layers are implemented in the gNB-CU (or en-gNB-CU) and RLC, MAC and physical layer protocol layers are implemented in the gNB unit-DU (or en-gNB-DU).

A centralized unit is connected to one or more distributed units via the F1 interface. The gNB-CU and gNB-DU entities can be co-located on the same site or deployed at different sites.

Co-location allows a swap of 4G base stations to 5G and ensures compatibility with the eNB entity. Deployment at relocation sites is intended for the virtualization of radio access (C-RAN: Cloud RAN) on COTS (Commercial Off The Shelf) server blades called LW-NFVI (Light-weight Network Function Virtualization Infrastructure) managed by an RIC (RAN Intelligent Controller).

Management of virtual machines is carried out by an intelligent controller (NFVM: Network Function Virtualization Management), which is the responsibility of an orchestrator (MANO: NFV MANagement and Orchestration).

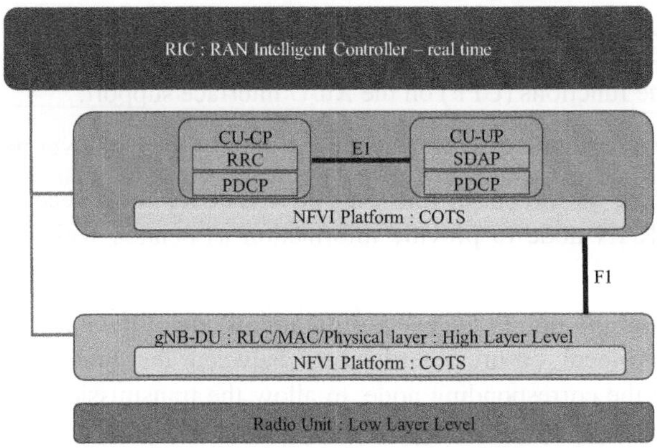

Figure 2.6. *Virtualization of radio access*

The NG-RAN architecture offers two possible deployments: a non-centralized architecture and a centralized architecture.

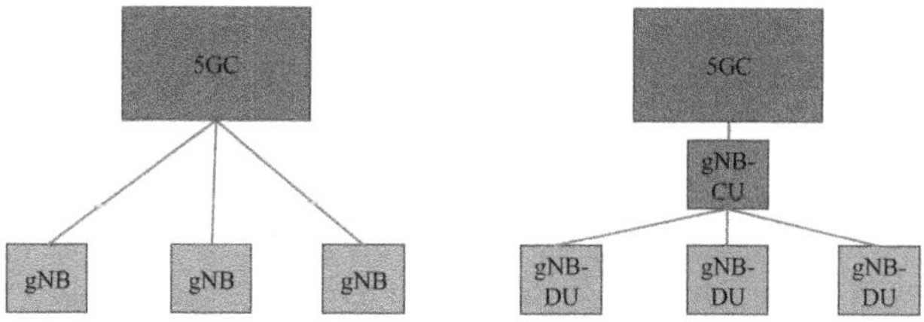

Figure 2.7. *NG-RAN architecture: distributed and centralized*

The non-centralized architecture allows evolution continuity of the radio access network through a swap. Each gNB entity can interconnect other gNB or ng-eNB entities via the Xn interface.

The centralized architecture allows the implementation of the CoMP (Cooperation Multi-Point) coordination mechanism by synchronously controlling several gNB-DUs.

The functional decomposition between gNB-DU and gNB-CU is also flexible. The eCPRI protocol offers several possible decompositions:

– option 1 is similar to the 5G-NSA dual connectivity architecture. For this option, the RRC and SDAP protocols are hosted in the gNB-CU, while the PDCP, RLC, MAC and physical layer protocols are localized by the gNB-DU. This option allows centralization of the management of the radio interface via the RRC protocol, and the separation of the medium (split-bearer) is managed at the level of the distributed units;

– option 2 has similarities with the Xn interface for the user plane. For the control plane, the RRC and PDCP protocols are hosted at the centralized gNB-CU and the RLC, MAC and physical layer protocols are located at the gNB-DU level. This option allows traffic aggregation between 4G and 5G base stations;

– option 3 divides the RLC protocol into two parts: the RLC high layer manages the message acknowledgment function (ARQ) and the low function layer manages the segmentation of packets. The gNB-CU entity hosts the RRC and PDCP protocols and manages the higher functions of the RLC (RLC-high) protocols. The gNB-DU entity manages the lower functions of the lower RLC protocol (RLC-low), MAC and the physical layer. This option allows the aggregation of traffic between 4G radio access and 5G radio access and load balancing between the two nodes. RLC state information is saved at the upper RLC sublayer and managed with the UE context. Retransmission is also managed at the level of the gNB-CU; it is thus possible to increase the number of terminals connected to the gNB_DU;

– option 4: the RLC protocol and higher layers are located in the gNB-CU, while the MAC protocol, as well as the physical layer, are hosted in the gNB-DU;

– option 5 splits the MAC functions and groups the low layer MAC function with the physical layer on the gNB-DU. Thus, the H-ARQ protocol is localized by the gNB-DU, while the scheduler is centralized at the level of the gNB-CU, thus making it possible to improve real-time management of interference between cells (ICIC: Inter-Cell Interference Coordination);

– option 6: the MAC protocol and the upper layers are hosted in the gNB-CU, while the physical layer is located in the gNB-DU. As the MAC protocol is centralized, coordination between transmission points is facilitated (COMP JT Joint Transmission);

– option 7 splits the physical layer into two parts: physical functionalities of the lower layer are implemented at the level of the gNB-DU; the other radio functionalities of the upper layer, the MAC, RLC, PDCP and RRC protocols are hosted at the gNB-CU level.

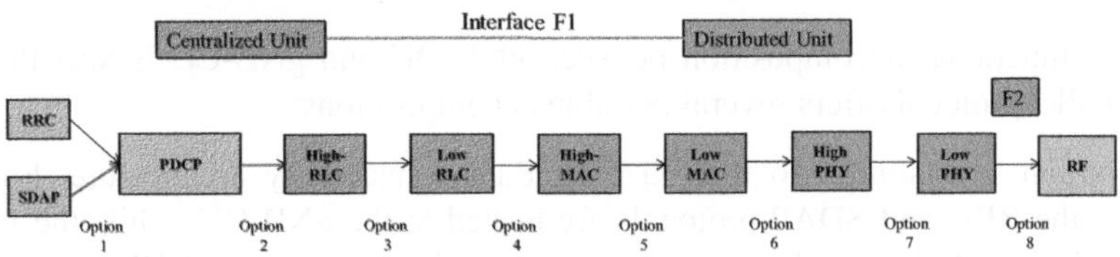

Figure 2.8. *gNB-CU and gNB-DU functional decomposition options*

The main goal of gNB decomposition is to reduce the throughput on the eCPRI link. High layer protocols host in the gNB-CU and low layer protocols host in the gNB-DU. The radio signal is reconstructed at the level of the active antenna (F2 interface).

The functionalities of the physical layer are split into several primary physical functions to transform the transport block transmitted by the MAC layer into an RF signal transmitted at the level of the antennas.

Option 7 is divided into three variants:

– variant 7.1: the lower part hosted in the gNB-DU includes the inverse fast Fourier transform (IFFT) for both downlink and uplink, while the higher part hosted in the en-gNB-CU contains other functions;

– variant 7.2: the lower part hosted in the en-gNB-DU adds the mapping function to the resource elements for both downlink and uplink;

– variant 7.3: channel coding is performed at the gNB-CU level and lower functionalities are implemented at the gNB-DU level.

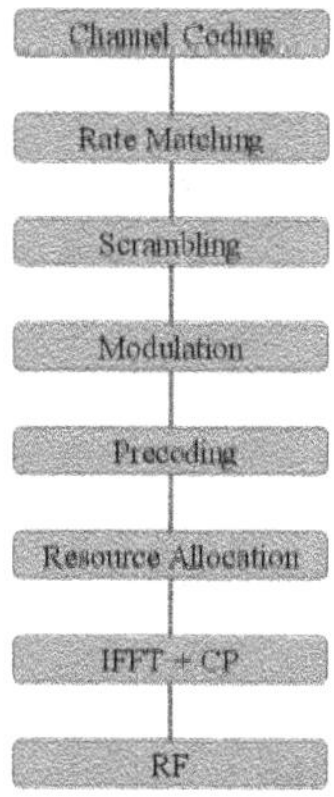

Figure 2.9. *Functions of the physical layer*

2.1.3.1. *Control plane function F1-C*

F1-C control functions support:

– management functions of the F1 interface: configuration or update of the gNB-CU, configuration or update of the gNB-DU, synchronization, error detection and restarting of the gNB unit;

– coordination of radio resource sharing (RAN Sharing);

– exchanges of broadcast information messages. NR-MIB and SIB1 messages are the responsibility of the gNB-DU and the other messages are handled by the gNB-CU;

– management of user contexts. This function is responsible for establishing or modifying a user context for data exchange (DRB) or signaling (SRB). The establishment of the F1 UE context is initiated by the gNB-CU. The gNB-DU can accept or decline the establishment request depending on the resources available (gNB-DU admission criteria). Modification of the F1 UE context can be initiated either by the gNB-CU or by the gNB-DU;

– exchanges of RRC messages between gNB-DUs and gNB-CUs.

The gNB-DU is responsible for scheduling and broadcasting information messages.

The scheduling is based on context UE parameters which are provided from the gNB-CU to the gNB-DU during the procedure of creation of context (UE Context Setup Procedure) or during the procedure of modification of context (UE Context Modification Procedure). The user context contains the QoS rules for the radio bearer and the S-NSSAI (Single Network Slice Assistance Indicator) for the network slice.

The gNB-CU performs the mapping between the PDU session streams and the radio bearers through the SDAP layer. The gNB-CU provides a radio QoS profile to the gNB-DU (DRB QoS Profile).

The Carrier Aggregation (CA) packet duplication function is handled by the gNB-CU. The gNB-CU controls the establishment of two GTP-U tunnels with the gNB-DU and instructs the gNB-DU to reconfigure the special secondary radio cell (SPcell).

2.1.3.2. *User plane functions F1-U*

User plane functions F1-U support:

– transfer of user data;

– flow control. This function controls the transmission of user data in the donwlink direction to the gNB-DU;

– inter-gNB-DU mobility when the user moves between two gNB-DUs controlled by the same gNB-CU;

– intra-gNB-DU inter-cell user mobility when the user moves between two cells of the same gNB-DU. The gNB-CU sends the UE context modification procedure to the mobile;

– NE-DC mobility: the user moves from an ng-eNB-DU to a gNB-DU controlled by the same gNB-CU.

2.1.4. *Protocol stack on the NG interface*

The N1 interface is the interface between the AMF function and the mobile. It supports signaling carried by the NAS (Non-Access Stratum) protocol.

The N2 interface is the interface between the NG-RAN radio node and the AMF function.

NAS messages, exchanged between the mobile and the AMF function, are transported by RRC messages on the Uu interface and NG-AP (Next Generation Application Protocol) messages on the N2 interface.

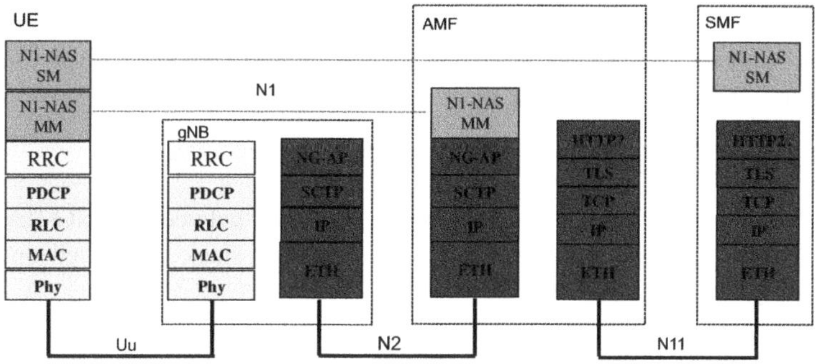

Figure 2.10. *Protocol stack of the control plane between the mobile and SMF function*

The NG-RAN access node establishes an association with the AMF function. Several associations are created in order to connect the access node to several AMF functions.

The functions of the N2 interface are established by NG-AP messages and are divided into two groups:

– generic functions intended to manage the association between the NG-RAN access node and the AMF function. These functions are independent of the mobiles;

– the functions associated with the mobile.

The NG-AP application layer protocol enables the exchange of control data between the NG-RAN radio access and the AMF function, via the N2 logical interface.

The NG-AP protocol is extended to the non-3GPP radio access network between the N3IWF (Non-3GPP InterWorking Function) and the AMF function.

The main functions associated with the mobile established through NG-AP messages concern:

– management of mobile attachment and monitoring its mobility (N1-NAS MM);

– management of the bearers established between the NG-RAN access node and the 5G core network (N1-NAS SM).

NAS signaling messages exchanged between the mobile and the AMF function are secured using an encryption key and an integrity key.

The N2 control plan separates the functions related to the AMF entity and the functions related to the SMF entity. For session request (SM) exchanges, the AMF transparently transfers the NG-AP messages to the SMF.

2.1.4.1. *NG-C interface*

The NG-C control plane interface is the interface between an NG-RAN node (gNB, ng-eNB) and the AMF function. The protocol stack is defined on the N2 interface (Figure 2.10).

From a user point of view, the functions on the NG-C control plane interface allow:

– management of the NG interface;

– UE context management (the UE context management functionality between the NG-RAN access and the 5G core network), including the context of the PDU session, the security key exchange, the mobility restriction list, the radio capacities and the security capacities of the mobile. Context management includes creating, modifying and deleting the context;

– management of UE mobility in connected mode (handover);

– notification of the RRC_INACTIVE state of the mobile to the AMF function;

– NAS message transport;

– NAS request routing (e.g. when registering on the dedicated network slice);

– session notification (paging);

– the management of PDU sessions for the establishment and deletion of PDU sessions, as well as for the transfer of a PDU session between two NG-RAN nodes via the core network (N2 handover).

From a system point of view, the functions of the NG-C interface allow us to:

– update the AMF configuration and NG-RAN radio nodes to ensure the interoperability of procedures on the NG-C interface;

– report errors which are not defined in specific messages;

– manage a 5G core network congestion state: the AMF function indicates to the NG-RAN the start or end of the core network congestion state;

– ensure bi-directional transfer of radio access configuration information to the AMF function (e.g. for a SON functionality);

– broadcast ETWS (Earthquake and Tsunami Warning System) type safety messages;

– re-initialize the radio access node.

2.1.4.2. *NG-U interface*

The NG-U user plane interface is the interface between the network node and the UPF.

The NG-U interface provides unreliable PDU data transfer services between the radio access node and the UPF.

2.2. Procedures on the radio network access

2.2.1. *XnAP procedures*

XnAP procedures consist of elementary procedures (EP) which can be of two types:

– class 1: request with response (failure or success, see Table 2.1);

– class 2: request without response (Table 2.2).

Elementary procedure – class 1	Request
Handover Preparation	HANDOVER REQUEST
Retrieve UE Context	RETRIEVE UE CONTEXT REQUEST
S-NG-RAN node Addition Preparation	S-NODE ADDITION REQUEST
M-NG-RAN node initiated S-NG-RAN node Modification Preparation	S-NODE MODIFICATION REQUEST
S-NG-RAN node initiated S-NG-RAN node Modification	S-NODE MODIFICATION REQUIRED
S-NG-RAN node initiated S-NG-RAN node CHANGE	S-NODE CHANGE REQUIRED
M-NG-RAN node initiated S-NG-RAN node Release	S-NODE RELEASE REQUEST
S-NG-RAN node initiated S-NG-RAN node Release	S-NODE RELEASE REQUIRED
Xn Setup	XN SETUP REQUEST
NG-RAN node Configuration Update	NG-RAN NODE CONFIGURATION UPDATE
Cell Activation	CELL ACTIVATION REQUEST
Reset	RESET REQUEST
Xn Removal	Xn REMOVAL REQUEST

Table 2.1. *XnAP elementary procedures – Class 1*

Elementary procedure – class 2
Handover Cancel
SN Status Transfer
RAN Paging
Xn-U Address Indication
S-NG-RAN node Reconfiguration Completion
S-NG-RAN node Counter Check
UE Context Release
RRC Transfer
Error Indication
Notification Control Indication
Activity Notification
Secondary RAT Data Usage Report

Table 2.2. *XnAP elementary procedures – Class 2*

Three main procedures occur on the Xn interface:

– basic mobility procedures for managing mobile mobility within the NG-RAN access network;

– dual connectivity procedures;

– global procedures which involve two NG-RAN nodes.

2.2.1.1. *Procedures for mobility management*

The mobility procedures are used to manage the mobile when it is in the connected state (RRC_CONNECTED) or inactive state (RRC_INACTIVE).

Mobility procedures are implemented for the following tasks:

– handover management;

– preparation of handover initiated by the source NG-RAN;

– UE context release initiated by the target NG-RAN;

– cancellation of handover;

– transfer of SN PDCP status;

– transfer of UE context for reestablishment of an RRC connection when UE move to a gNB that is different from the one with which it suspended its RRC connection to go to the RRC_INACTIVE state;

– management of RAN Paging when the 5G core network transmits IP flows to the gNB on which the terminal has suspended its RRC connection before switching to the RRC_INACTIVE mode.

2.2.1.2. *Dual connectivity*

The objectives of dual connectivity procedures are to add, modify or release the resources of the secondary base station:

– adding a UE context to the secondary base station SN initiated by the master base station MN;

– release of the UE context at the secondary base station SN. This request is initiated by the master base station MN;

– change the UE context;

– RRC message transfer between the MN entity and the terminal via the SN (Split Bearer) entity;

– transfer of activity notifications in order to inform, in real time, the master node (MN) of the user plane resources that are used by the secondary station (SN);

– transfer of the data consumption measurement report between the secondary node SN and master node MN;

– request for the value of the SN PDCP sequence number made by the secondary node (SN) to the master node (MN) in order to associate this value with the media established at the SN.

2.2.1.3. *Global procedures*

Global procedures allow:

– establishment or deletion of the Xn connection. This procedure enables the exchange of configuration in order that two neighbor NG-RAN nodes can communicate on the Xn-C interface. The information transmitted relates to the list of AMF entities in the region concerned, the list of cells served by the node and security information;

– updating the configuration of the nodes (e.g. updating the TAI lists);

– deactivation of cells in the event of inactivity of a neighboring cell to reduce the carbon impact;

– cell activation;

– error notification when an error is detected on a received message, a message for which there is no failure response;

– re-initialization of a node.

2.2.2. *F1 interface procedures*

F1 Application Protocol (F1AP) provides signaling services between a gNB-CU and a gNB-DU. There are two types of services, the services associated with a UE and the services associated with the F1 interface (not associated with a UE).

The F1AP protocols consist of elementary procedures (EP) which can be of two types:

– class 1: request with response (failure or success, see Table 2.3);

– class 2: request without response (Table 2.4).

Elementary procedure – class 1	Request
Reset	RESET
F1 Setup	F1 SETUP REQUEST
gNB-DU Configuration Update	GNB-DU CONFIGURATION UPDATE
gNB-CU Configuration Update	GNB-CU CONFIGURATION UPDATE
UE Context Setup	UE CONTEXT SETUP REQUEST
UE Context Release (gNB-CU initiated)	UE CONTEXT RELEASE COMMAND
UE Context Modification (gNB-CU initiated)	UE CONTEXT MODIFICATION REQUEST
UE Context Modification Required (gNBDU initiated)	UE CONTEXT MODIFICATION REQUIRED
Xn Setup	XN SETUP REQUEST
Write-Replace Warning	WRITE-REPLACE WARNING REQUEST
PWS Cancel	PWS CANCEL REQUEST
GNB-DU RESOURCE COORDINATION	GNB-DU RESOURCE COORDINATION REQUEST

Table 2.3. *F1 elementary procedures – Class 1*

Elementary procedure – class 2
Error Indication
UE Context Release Request (gNBDU initiated)
Initial UL RRC Message Transfer
DL RRC Message Transfer
UL RRC Message Transfer
UE Inactivity Notification
System Information Delivery
Paging
Notify
PWS Restart Indication
PWS Failure Indication

Table 2.4. *F1 elementary procedures – Class 2*

The procedures on the F1 interface concern:

– F1 interface management procedures;

– management procedures of the UE context;

– RRC message transfer procedures;

– paging procedures;

– information system broadcasting.

2.2.2.1. *The F1 interface management procedure*

The reset procedure is used to initialize or re-initialize the UE contexts following a failure of the gNB-DU or the gNB-CU.

2.2.2.2. *Dual connectivity*

The objectives of dual connectivity procedures are to add, modify and release the resources of the secondary base station:

– adding a context to the secondary base station SN initiated by the master base station MN;

– release the UE context at the secondary base station SN. This request is initiated by the master base station MN;

– UE context updating;

– RRC message transfer between the MN entity and the terminal via the SN (Split Bearer) entity;

– transfer of activity notifications in order to inform the master node (MN) in real time of the user plane resources that are used at the secondary station (SN);

– transfer of the data consumption measurement report between the secondary node and the master node;

– request for the value of the SN PDCP sequence number made by the secondary node to the master node in order to associate this value with the media established at the secondary SN node.

2.2.2.3. *Global procedures*

Global procedures allow:

– establishment or release of the Xn interface. This procedure enables the exchange of configuration information so that two neighbor NG-RAN nodes can communicate on the Xn-C interface. The information transmitted concerns the list of AMF entities in the region concerned, the list of cells served by the node and security information;

– updating the configuration of the nodes such as updating the TAI lists;

– deactivation of cells in the event of inactivity of a neighboring cell to reduce the carbon impact;

– cell activation;

– error notification when an error is detected on a received message, a message for which there is no failure response;

– re-initialization of an NG-RAN node.

2.2.3. *NG-AP procedures*

The 5G core network provides data transmission services by establishing tunnels between the mobile and a data network (DN).

PDU sessions are initiated by the mobile and are controlled by the AMF function.

To set up a PDU session, the mobile first establishes a signaling radio bearer (SRB) with gNB to transmit NAS messages towards the core network. The user plane is on the control of the core network from gNB to the data network (DN).

Connection management includes the establishment and release of signaling between the mobile and the AMF entity through the N1 interface (Figure 2.10). The N1 interface is based on the Uu interface (between the mobile and the radio access) and on the N2 interface between the radio node and the AMF function.

The SM-NAS message (Session Management NAS) controls the establishment of the session. This message is transmitted between the mobile and the SMF function. The content of the SM-NAS message is transparent to the AMF function.

The control plane allows the establishment of a user plane between the mobile and the 5G core network. The established tunnel can be an IP connection (IPv4, Ipv6) or a point-to-point Ethernet connection.

The GTP-U encapsulation protocol allows multiple PDU sessions to be multiplexed between the radio node and the UPF, via the N3 interface.

NG-AP protocols consist of elementary procedures (EP) which can be of two types:

– class 1: request with response (failure or success, see Table 2.5);

– class 2: request without response (Table 2.6).

Elementary procedure – class 1	Request
AMF Configuration Update	AMF CONFIGURATION UPDATE
RAN Configuration Update	RAN CONFIGURATION UPDATE
Handover Cancellation	HANDOVER CANCEL
Handover Preparation	HANDOVER REQUIRED
Handover Resource Allocation	HANDOVER REQUEST
Initial Context Setup	INITIAL CONTEXT SETUP REQUEST
NG Reset	NG RESET
NG Setup	NG SETUP REQUEST
Path Switch Request	PATH SWITCH REQUEST
PDU Session Resource Modify	PDU SESSION RESOURCE MODIFY REQUEST
PDU Session Resource Modify Indication	PDU SESSION RESOURCE MODIFY INDICATION
PDU Session Resource Release	PDU SESSION RESOURCE RELEASE COMMAND
PDU Session Resource Setup	PDU SESSION RESOURCE SETUP REQUEST
UE Context Modification	UE CONTEXT MODIFICATION REQUEST
UE Context Release	UE CONTEXT RELEASE COMMAND
Write-Replace Warning	WRITE-REPLACE WARNING REQUEST
PWS Cancel	PWS CANCEL REQUEST
UE Radio Capability Check	UE RADIO CAPABILITY CHECK REQUEST

Table 2.5. *NG-AP elementary procedures – Class 1*

Elementary procedure – class 2
Downlink RAN Configuration Transfer
Downlink RAN Status Transfer
Downlink NAS Transport
Error Indication
Uplink RAN Configuration Transfer
Uplink RAN Status Transfer
Handover Notification
Initial UE Message
NAS Non-Delivery Indication
Paging
PDU Session Resource Notify
Reroute NAS Request
UE Context Release Request
Uplink NAS Transport
AMF Status Indication
PWS Restart Indication
PWS Failure Indication
UE Radio Capability Info Indication
RRC Inactive Transition Report
Overload Start
Overload Stop

Table 2.6. *NG-AP elementary procedures – Class 2*

The procedures on the NG interface concern:

– PDU session management procedures;

– the procedures for managing the context of the UE;

– the mobility procedures of the UE;

– notification procedures (paging);

– NAS message transport procedures;

– global procedures which involve an NG-RAN node and a core network function.

2.2.3.1. *PDU session management procedures*

The procedures related to PDU sessions allow resources to be allocated on the NG-U interface:

– request for establishment of a PDU session from the AMF to a radio access node;

– modification of a PDU session from the AMF to a radio access node;

– release of a PDU session from the AMF to a radio access node;

– notification from an NG-RAN node that some QoS flows of a PDU session are about to be released.

2.2.3.2. *UE context management procedures*

In order to establish a PDU session, it is necessary to first secure the radio signaling bearer and data bearer exchanged on the Uu interface.

The NG-RAN entity establishes a context for each UE, comprising:

– security keys;

– the radio capacities of the terminal and its security capacities;

– the PDU session context;

– the mobility restriction list.

The creation of the context or the modification of the context is triggered by the AMF function.

The context release is triggered either by the AMF or by the NG-RAN node (e.g. upon completion of a handover).

2.2.3.3. *Mobility procedures*

AMF handles the mobility of UE. For each registered UE, AMF establishes a context for which the mobile is either in the idle state (CM_IDLE) or in the connected state (CM_CONNECTED).

In terms of the NG-RAN node, the state of the mobile is either in RRC_IDLE or RRC_CONNECTED or RRC_Inactive.

The state mapping between the NG-RAN node and AMF is:

– RRC_IDLE state and CM_IDLE;

– RRC_CONNECTED state and CM_CONNECTED state;

– RRC_INACTIVE state and CM_CONNECTED connected state.

When the mobile is in the idle state, the location of the terminal at the core network level is defined by the TAI (Tracking Area Identity).

The TAI corresponds to the coverage of a set of NG-RAN nodes. The TAI value is saved in the UE context of the mobile and in the UE context established at the AMF function.

When the mobile is in the RRC_CONNECTED state, the AMF function knows the NCGI (NR Cell Global Identifier) identity of the cell on which the mobile is exchanging data.

The base station manages the allocation of radio resources and can trigger session continuity with another base station according to measurement reports from the UE.

The handover mechanism is initiated from the source base station to a target base station. If the target base station does not have an Xn interface with the source base station, then the handover request is transmitted to the AMF via the Handover Required request.

The AMF function requests the target base station to reserve radio resources to the mobile by the Handover Request. The target base station informs the AMF of the execution of the handover via the Handover Notify request. The completion of the handover is always finalized by a modification of the routing table of the UPF. The target base station sends the tunnel information via the PATH SWITCH REQUEST.

When the mobile is in the RRC_INACTIVE state, the core network does not manage its mobility, the AMF function can nevertheless subscribe to a service allowing it to be notified when the terminal changes from the RRC_INACTIVE state to the RRC_CONNECTED state.

2.2.3.4. *Notification procedures*

The notification procedure is initiated by the AMF function and aims to locate the mobile on an NG-RAN node. The AMF node knows the TAI location of the mobile and the value of the DRX (Discontinuous Reception) counter if this is activated.

When the mobile is in the CM_IDLE state, the AMF function requests the NG-RAN nodes of the TAI zone of the mobile to broadcast a notification message (paging).

In the event of base station congestion, the base station processes priority notifications. The priority is defined in an optional field of the PAGING message sent by the AMF to the gNB base station.

2.2.3.5. *NAS message transport procedures*

UE can transmit data messages like SMS or small amounts of data for IoT terminals in a NAS request.

For uplink transmission, the NG-RAN node transfers the message encapsulated in the RRC protocol to the AMF, via either the Initial UE Message for the first transmission or the Uplink NAS Transport message for subsequent packets.

For downlink transmission, the AMF sends the Downlink NAS Transport request with possibly a TAI restriction list and a paging priority indicator.

If the NG-RAN node does not transmit the message, it informs the AMF via the NAS Non-Delivery Indication message.

When a mobile makes an attachment request, the NG-RAN node transfers the message to a source AMF function. In the case of virtualization of network functions, and slicing of the network (Network Slicing), the choice of the radio node is defined on the basis of the information requested by NSSAI.

2.2.3.6. *Global procedures*

Global procedures allow us to:

– configure the NG-RAN nodes and the AMF function in order to ensure the full interoperability of the two entities (NG Setup message);

– reinitialize the NG-RAN node in the event of a fault at the level of the radio node or of the 5GC core network via the NG Reset message;

– update the configuration of the NG-RAN node and the configuration of the AMF function;

– inform the radio node in the event of the start or end of AMF congestion.

2.3. Identities of the XnAP and NG-AP application protocols

An application protocol identity (AP ID) is allocated per mobile for each new logical connection creation on the NG-RAN interfaces:

– RAN UE NGAP ID uniquely identifies for a gNB, the UE connection on the NG interface. The identity is created by the base station and is transmitted to the AMF entity;

– AMF UE NGAP ID uniquely identifies the connection of the mobile on the NG interface for an AMF. The identity is created by the AMF function and is transmitted to the base station;

– M-NG-RAN node UE XnAP ID uniquely identifies the mobile on the Xn interface in the case of dual connectivity. The identifier is created by the master base station and is transmitted to the secondary base station;

– S-NG-RAN node UE XnAP ID uniquely identifies the mobile on the Xn interface in the case of dual connectivity. The identifier is created by the secondary base station and is transmitted to the master base station;

– gNB-CU UE F1AP ID uniquely identifies the mobile on the F1 interface. The identifier is created by the gNB-CU entity;

– gNB-DU UE F1AP ID uniquely identifies the mobile on the F1 interface. The identifier is created by the gNB-DU entity;

– gNB-CU UE E1AP ID uniquely identifies the mobile on the E1 interface. The identifier is created by the gNB-CU entity;

– gNB-DU UE E1AP ID uniquely identifies the mobile on the E1 interface. The identifier is created by the gNB-DU entity.

2.4. References

All standards can be downloaded from the ETSI website: https://www.etsi.org/standards.

3GPP TS 29.413
Application of the NG Application Protocol (NGAP) to non-3GPP access
Version 15.0.0 Release 15

3GPP TS 37.340
Universal Mobile Telecommunications System (UMTS); LTE; 5G; NR; Multi-connectivity; Overall description; Stage-2
Version 15.5.0 Release 15

3GPP TS 38.401
5G – NG-RAN: Architecture description;
Version 15.5.0 Release 15

3GPP TS 38.410
5G – NG-RAN: NG general aspects and principles;
Version 15.5.0 Release 15

3GPP TS 38.413
5G – NG-RAN: NG Application Protocol (NGAP);
Version 15.5.0 Release 15

3GPP TS 38.423
5G – NG-RAN: Xn Application Protocol (XnAP);
Version 15.1.0 Release 15

3GPP TS 38.470
5G – NG-RAN: F1 general aspects and principles;
Version 15.7.0 Release 15

3

NG-RAN Network – Procedures

3.1. General procedure of the 5G-NSA mode

The initial deployment of the 5G service is based on 5G-NSA option 3, also called dual EN-DC connectivity. When the phone is switched on, the mobile starts with a cell search procedure on the 4G/LTE (Long-Term Evolution) radio network. Once synchronized with the eNB (evolved Node Base station), the mobile reads the MIB (Master Information Block) system information carried by the BCCH (Broadcast Control Channel) beacon channel and the SIB (System Information Block) information carried by the PDSCH (Physical Downlink Data Channel) physical channel, in order to acquire the radio parameters of the cell.

After synchronization, the mobile requests a signaling radio bearer to the eNB. A signaling radio bearer (SRB) is used for the transmission of RRC and NAS messages. The establishment of a RRC connection starts via the random access procedure. The procedure is initiated by the mobile and the radio link is controlled by the base station.

During the UE attach, the eNB transmits a radio configuration to the UE. The radio configuration is used to define the type of measurements to be performed on the LTE and NR (New Radio) radio accesses for dual connectivity.

When the UE requests dedicated SRB and DRB (Data Radio Bearer) to exchange message/data with the network, the eNB applies a RRC Connection reconfiguration procedure. This procedure aims to establish a UE context to define a radio identification in the connected mode (C-RNTI: Connected Radio Network Temporary Identifier), and an E-RAB (E-UTRAN Radio Access Bearer) for one-to-one mapping between the E-RAB and EPS bearer. For dual connectivity, two radio bearers are set up, one between the mobile and the master MN (Master Node) and the other between the mobile and the secondary SN (Secondary Node).

In the case of the 5G-NSA option 3x, the MN node is the eNB (evolved Node Base Station) entity and the SN node is the en-gNB (E-UTRAN/NR new Generation Base Station) entity. A traffic tunnel is set up between the en-gNB entity and the 4G EPC (Evolved Packet Core) network.

3.1.1. *LTE search procedure*

The selection of the 4G radio access network is initiated by the mobile. When the mobile switches on, it searches for the synchronization signal. The synchronization signal is periodically transmitted by each eNB.

If the mobile was previously attached to a PLMN (Public Land Mobile Network), it listens to the frequency that was used (and stored on the SIM card) before detachment.

If the mobile is switched on for the first time, or in the case of roaming (first roaming connection), the mobile performs a frequency scan on the possible LTE bands (depending on the model of the mobile).

The synchronization signal identifies the cell with the PCI (Physical layer Cell Identity). The PCI of the 4G cell is calculated from two values: $N_{ID}^{(2)}$ and $N_{ID}^{(1)}$, which are carried respectively by the primary synchronization signal PSS (Primary Synchronization Signal) and the secondary synchronization signal SSS (Secondary SS).

The number $N_{\mathrm{ID}}^{\mathrm{cell}}$ for the Physical layer Cell Identity (PCI) is equal to $N_{\mathrm{ID}}^{\mathrm{cell}} = 3N_{\mathrm{ID}}^{(1)} + N_{\mathrm{ID}}^{(2)}$.

$N_{\mathrm{ID}}^{(1)}$ represents the group number and can take values ranging from 0 to 167. It is determined by the Secondary Synchronization Signal (SSS).

$N_{\mathrm{ID}}^{(2)}$ represents the number in the group and can take values ranging from 0 to 2.

The Primary Synchronization Signal (PSS) performs the following functions:

– frequency synchronization;

– time synchronization, at the level of the Orthogonal Frequency Division Multiplexing (OFDM) symbol, the time slot, the subframe (1 ms periodicity) and the half-frame (5 ms periodicity);

– value determination for the number $N_{\mathrm{ID}}^{(2)}$.

The Secondary Synchronization Signal (SSS) performs the following functions:

– time synchronization at the frame level (10 ms periodicity);

– determination of the duplex mode: either Frequency Division Duplex (FDD) mode or Time Division Duplex (TDD) mode;

– determination of the type of Cyclic Prefix (CP) to be used, normal or extended;

– determination of the value of number $N_{\mathrm{ID}}^{(1)}$.

The MIB message is carried by the Physical Broadcast Channel (PBCH). The MIB is a information message that is 24 bits, which provides:

– the bandwidth of the radio channel, for the downlink direction;

– the System Frame Number (SFN);

– the configuration of the Physical HARQ Indication Channel (PHICH);

– the repetition period of the SIB1 message (System Information Block Type 1).

The SIB1 message transmits the identity of the serving PLMN (PLMN ID, TA Code, Cell Id), the cell selection criteria (minimum threshold) and the radio measurement configuration to be carried out in the standby mode (RRC_IDLE). The PLMN ID allows the mobile to select the PLMN based on the priority data configured in the USIM (Universal Subscriber Identity Module).

The USIM module contains a list of:

– preferred PLMNs sorted in order of priority;

– equivalent PLMNs (optional).

The SIB2 message provides information relating to the configuration of the radio resources allocated to the physical PRACH (Physical Random Channel).

According to the power received from each serving cell of the PLMN network, the mobile selects the best cell. The power measurement is calculated from Cell Specific Reference Signals (CRS).

The previously calculated value N_{ID}^{cell} also determines the mapping of the Cell-Specific RS physical signal to the Resource Element (RE).

The RSRP (Reference Signal Received Power) measurement is the average power measured on the first antenna port. The RSRQ (Reference Signal Received Quality) power measurement compares the power received from the cell (RSRP) to the total power of the received signal (RSSI- Reference Signal Strength Indicator). The RSSI corresponds to the power of the signal received from the cell, the noise and interference present is added by averaging the measurement on N block resources.

Cell re-selection is also based on the PLMN priority list stored in the USIM module. The mobile selects a (new) serving cell by comparing the measurement made at the physical level (L1) between all of the cells that belong to the list of neighbor cells. The re-selection threshold used for comparison and the list of neighbor cell are carried in an SIB message transmitted by the serving eNB entity. As long as the RSRP power measurement of the serving cell minus an offset is greater than or equal to the power measured at the level of the neighboring cell, the mobile remains on the serving cell.

The re-selection criteria are defined for each type of radio access (4G intra and inter frequencies, 3G and 2G).

3.1.2. *Random access procedure*

The random access procedure is initialized by the mobile and it is required in the following cases:

– when establishing the connection to the eNB entity;

– when changing the cell during the session (handover);

– when updating Timing Advance (TA);

– when re-establishing the connection to the eNB entity.

The random access procedure is said to be with contention when the mobile randomly chooses the resources (PRACH channel, preamble). A contention may occur for radio connection establishment or reestablishment.

The random access procedure is said to be without contention when the eNB entity is providing the resource to be used to the mobile. This occurs for the handover or for the updating of the timing advance or for dual connectivity.

3.1.2.1. *Random access with contention*

The random access procedure with contention, during the establishment or reestablishment of the eNB entity connection, is described in Figure 3.1:

1) The mobile transmits the preamble in the PRACH physical channel; this preamble is randomly chosen in a list communicated by the eNB entity in the SIB2 system information.

Without the response of the eNB, the mobile retransmits the preamble and increases the transmission power. The maximum number of retransmission is limited by SIB2 information or by the RRC *ConnectionReconfiguration* message.

The risk of contention increases with the number of mobiles that try to access the same PRACH physical channel.

2) When the eNB entity receives the PRACH physical channel, it calculates the timing advance and transmits it to the mobile with the following information:

– the Downlink Control Information (DCI) in the PDCCH physical channel, in format 1A or 1C scrambled with RA-RNTI identity;

– the Random Access Response (MAC RAR) frame containing the index of the preamble, the timing advance, the resource to use (UL Grant) for the transmission in the Physical Uplink Shared Channel (PUSCH) and the Temporary Cell RNTI (TC-RNTI) in the PDSCH physical channel.

This identity is temporary because in case of contention, several mobiles may assume that this identity is allocated to them, thereby leading to contention.

The RA-RNTI identity is calculated as follows:

– $\mathrm{RA-RNTI} = 1 + t\_id + 10 * f\_id$;

– t_id is the index of the first subframe used by the PRACH physical channel ($0 \leq t\_id \leq 9$);

– f_id is the index of the PRACH physical channel, in the frequency domain ($0 \leq f\_id < 6$).

3) The mobile initializes its timing advance and responds with the RRC *ConnectionRequest* message containing:

– the Shortened Temporary Mobile Subscriber Identity (S-TMSI), if the mobile is already connected;

– a 40 bits random number, otherwise.

4) When the eNB entity receives the RRC *ConnectionRequest* message, it transmits to the mobile:

– the Downlink Control Information (DCI) information in the PDCCH physical channel;

– the heading MAC RAR containing the UE CRI (Contention Resolution Identity) control element. This control element reproduces the identity of the RRC *ConnectionRequest* message to resolve the contention issue.

The mobile extracts DCI information from the TC-RNTI identity. The mobile reads the description of its data in the PDSCH physical channel.

The TC-RNTI temporary identity becomes the C-RNTI identity allocated to the mobile.

5) The mobile shows its C-RNTI identifier in the control element of the MAC frame and confirms the connection through the RRC *ConnectionSetupComplete* message.

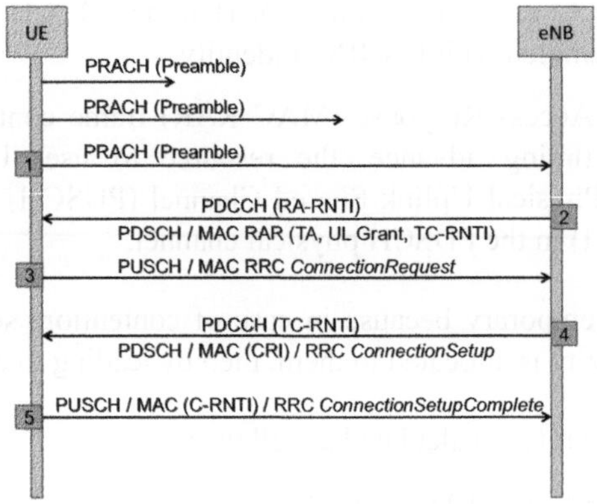

Figure 3.1. *Random access with contention*

3.1.2.2. *Random access without contention*

The random access procedure without contention, i.e. when changing the cell during the session, is described in Figure 3.2.

The mobile takes measurements on its cell and neighboring cells and transmits the results to the eNB entity in the RRC *MeasurementReport* message:

1) During the handover procedure based on the X2 interface, the source eNB entity provides the characteristics of the radio interface to the target eNB entity in a X2-AP HANDOVER REQUEST message.

2) The target eNB entity reserves the resources and responds to the source eNB entity with the X2-AP HANDOVER REQUEST ACK message.

This message contains the information element *Handover Command*, which specifies the preamble the mobile must transmit to the eNB entity for the random access procedure.

3) The source eNB entity forwards the information element *Handover Command* to the mobile in the RRC *ConnectionReconfiguration* message, which also triggers the handover.

4) 5) The random access procedure comprises the preamble transmission transmitted by the mobile and the MAC RAR frame response of the eNB entity.

6) The mobile connection is finalized when the mobile transmits the RRC *ConnectionReconfigurationComplete* message.

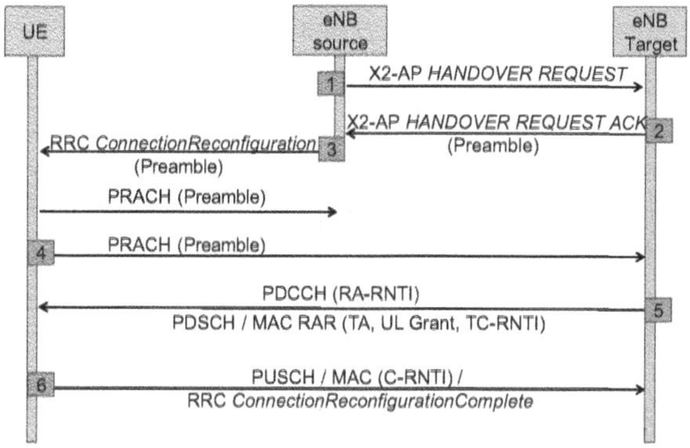

Figure 3.2. *Random access without contention, in case of handover*

3.1.3. *Data transfer*

During the attachment procedure, the 4G core network is aware of either of the mobile's support dual 4G and 5G connectivities (radio capabilities).

The Mobility and Management Entity (MME) performs the attachment procedure in order to authenticate the mobile. It checks with the HSS (Home Subscriber Server) database whether the customer has subscribed to the 5G-NSA offer.

If applicable, when the mobile wishes to establish an end-to-end bearer (E2E), the MME entity will inform the eNB entity that the mobile is authorized to connect to the 5G-NR radio access.

The selection of the eNB base station is performed by the mobile according to the PLMN priorities stored on the USIM module. A 5G-NSA mobile will therefore search for an eNB base station that can implement dual connectivity, as a priority. This information is contained in the SIB1 channel broadcast by the base station eNB.

3.1.3.1. *Establishment session procedure on the 4G radio access network*

1) 2) The mobile performs the random access procedure with the selected serving eNB base station (i.e. the selected eNB when the mobile is in standby mode). The access procedure aims to create a C-RNTI identity, control radio resources on UpLink and allow us to adjust its Timing Advanced (TA).

3) The mobile sends its session establishment request to the eNB entity in an RRC *ConnectionRequest* message. In this request, the mobile sends its C-RNTI radio identity, its S-TMSI core network identity or a random number and the object of its request (session request: Mobile Originating Signaling).

4) The eNB entity allocates a radio dedicated signaling Radio Bearer (SRB1) to the mobile via the RRC *ConnectionSetup* message. The downlink (DL) and uplink (UL) radio resources are controlled by the base station. On receipt of the RRC *ConnectionSetup* message, the mobile will use the dedicated radio resources to transmit the content of the NAS signaling message.

5) The mobile notifies the base station of the establishment of the SRB1 signaling bearer via the RRC *ConnectionSetupComplete* message and transmits the NAS information (e.g. an attachment request, a service request, an SMS) in the *DedicatedInfoNAS* field of the RRC *ConnectionSetupComplete message.*

The eNB entity transmits the session establishment request to the MME entity.

The MME provides the following information to the eNB:

– E-RAB radio support parameters, including QoS parameters and tunnel termination with the SGW (Serving Gateway) entity;

– the radio capabilities of the mobile;

– the security context of the mobile to secure radio access (the eNB security key).

The eNB entity will activate the DRB bearer by setting up:

6) 7) a procedure to secure the radio link (securing the AS access stratum);

8) 9) a reconfiguration of the RRC connection. During this request, the eNB entity transmits the measurement configuration.

At the end of this procedure, the mobile can securely exchange data with the eNB entity. The eNB entity confirms the establishment of the DRB radio bearer to the MME entity so that the latter can finalize the establishment of the tunnel between the eNB entity and the SGW entity handling the 4G core network traffic.

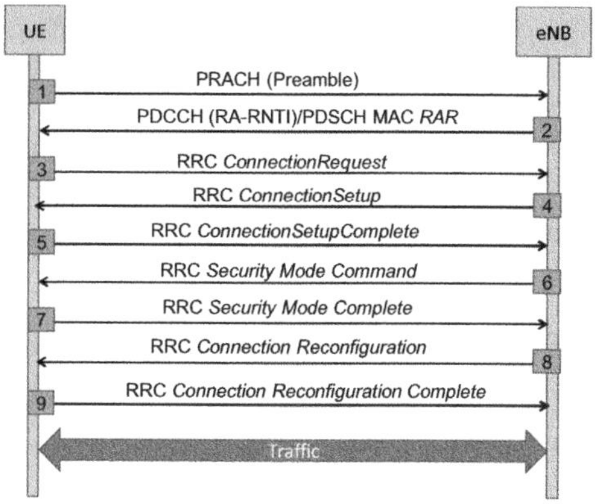

Figure 3.3. *Procedure for establishing a DRB*

3.1.3.2. *Establishment of dual connectivity option 3X*

The procedure for adding a secondary node (en-gNB entity) is initialized by the eNB entity (master node). It is used to establish context at the en-gNB entity to provide radio resources to the mobile.

For SCG (Secondary Cell Group), a new radio resource Pcell is required. The SCG is a group of serving cells assigned by the secondary cell, like the MCG (Master Cell Group) assigned by the master cell.

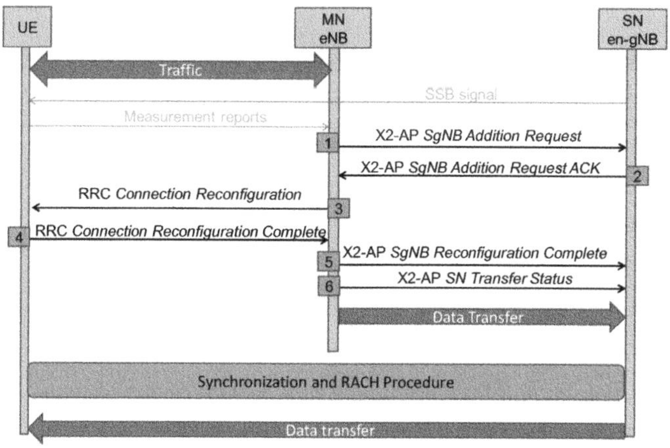

Figure 3.4. *Procedure to add a secondary node. For a color version of this figure, see www.iste.co.uk/launay/5g.zip*

1) The eNB entity sends the X2-AP *SgNB Addition Request* message to the en-gNB entity to allocate radio resources for the establishment of an S1-U bearer with the serving gateway (SGW). The X2-AP message allows the optional establishment of SRB3 signaling between the mobile and the en-gNB entity. The eNB entity sends the TNL transport layer information for the establishment of a tunnel, depending on the dual connectivity option (option 3, 3a or 3x) and the radio capabilities of the mobile.

At the secondary node, the RRM (Radio Resource Management) function, which manages the admission request, can accept or refuse the request of the master node.

2) If the request is accepted, the SN node (en-gNB entity) determines the primary cell and possibly the secondary cells (SCell). The radio resource configuration is sent to the eNB entity in a NR RRC configuration encapsulated in the X2-AP *SgNB Addition Request Acknowledge* message.

3) The eNB entity forwards the SN configuration to the mobile in the message RRC *ConnectionReconfiguration.*

4) The mobile applies the new configuration and acknowledges the MN (eNB entity) of the new allocation of the radio resource in the RRC *ConnectionReconfigurationComplete* message, including, if necessary, an NR RRC response message.

5) The MN (eNB entity) informs the SN (en-gNB entity) that the mobile has successfully completed the reconfiguration procedure, via the X2-AP *SgNB Reconfiguration Complete* message, eventually including the NR RRC response message.

6) In case of the EN-DC option 3x dual connectivity, the MN (eNB entity) only transmits the SN Status Transfer message for packets in the Radio Link Control Acknowledged mode (RLC-AM).

The mobile synchronizes on the MN (PCell of the en-gNB entity) and performs the random access procedure.

To establish the SI-U bearer between the en-gNB and SGW entities, the MN (eNB entity) transmits the S1-AP E-RAB *Modification Indication* message to the MME entity.

The SI-U bearer continues to set up, with GTPv2-C messages exchanged between the MME and SGW entities.

The MME entity acknowledges the MN (eNB entity) request by sending the S1-AP E-RAB Change Confirmation message.

The S1-U bearer between the SGW and en-gNB entities is complete.

3.1.3.3. *Changing a secondary node*

The procedure for changing a secondary node is initialized by either the MN (eNB entity – Figure 3.5) or the SN (en-gNB entity). It is used to transfer the mobile context of the source en-gNB entity to the target en-gNB entity.

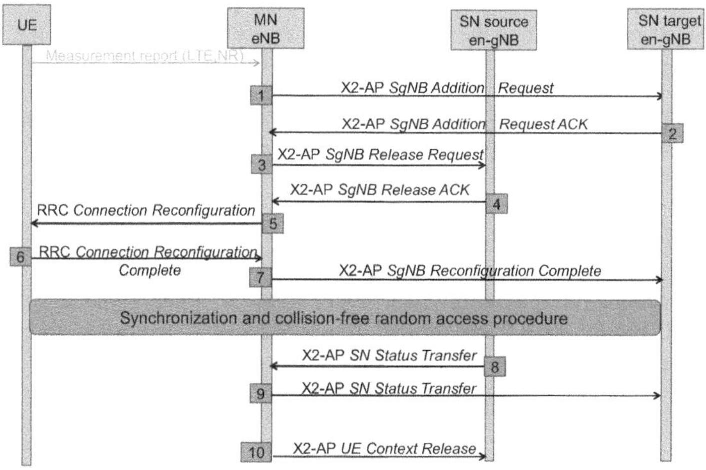

Figure 3.5. *Procedure initiated by the eNB to change the secondary node. For a color version of this figure, see www.iste.co.uk/launay/5g.zip*

1) 2) The procedure for changing the en-gNB entity starts with the X2-AP *SgNB Addition Request* and SgNB *Addition Request Acknowledge* message exchanges between the eNB and the target en-gNB entities.

3) 4) If the resource allocation from the target en-gNB entity is successful, the eNB entity triggers the resource release for the source en-gNB entity, with the exchange of the X2-AP *SgNB Release Request* and SgNB *Release Request Acknowledge* messages.

5) 6) The eNB entity triggers the new configuration at the mobile level in the RRC *ConnectionReconfiguration* message, containing the NR RRC configuration message generated by the target en-gNB entity. The mobile applies the new configuration and sends the RRC *ConnectionReconfigurationComplete* message.

7) The eNB entity informs the target en-gNB entity via the X2-AP *SgNB Reconfiguration Complete* message that the reconfiguration procedure is successful.

The mobile synchronizes on the PCell of the target en-gNB entity and performs the random access procedure without contention.

8) 9) If the bearer was configured in the acknowledgment mode, the source en-gNB entity sends the PDCP sequence number in the X2-AP SN *Status Transfer* message, that the eNB entity transfers to the target en-gNB entity.

The eNB entity requests the MME entity to establish the S1-U bearer between the target en-gNB entities and the SGW. The MME entity acknowledges the request and the SGW entity notifies the target en-gNB entity about the establishment of the downlink tunnel.

10) Upon receipt of the X2-AP UE *Context Release* message, the source en-gNB entity releases the resources allocated to the mobile.

3.1.4. *Removing a secondary node*

The procedure for removing a secondary node is either initialized by the MN (eNB entity, Figure 3.6) or by the SN (en-gNB entity). It is used to remove the mobile context at the SN (en-gNB entity).

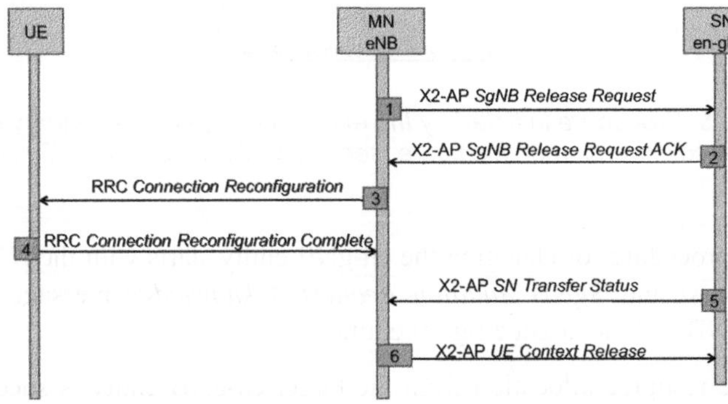

Figure 3.6. *Removing a secondary node initiated by the eNB entity*

1) The eNB entity initiates the procedure for removing the secondary node by sending the X2-AP *SgNB Release Request* message to the en-gNB entity. Data transfer to the eNB entity may be requested.

2) The en-gNB entity confirms the receipt by sending the X2-AP SgNB *Release Request Acknowledge* message.

3) 4) The eNB entity indicates in the RRC *ConnectionReconfiguration* message to the mobile that it must release the entire SCG bearer configuration. The mobile acknowledges the received message in the RRC *ConnectionReconfigurationComplete* response message.

5) If the released bearer uses the acknowledgment mode for RLC, the en-gNB entity sends the X2-AP SN *Status Transfer* message to avoid the loss of the packet.

The S1-U bearer establishment procedure between the eNB and SGW entities is set up.

6) Upon receipt of the X2-AP UE *Context Release* message, the en-gNB entity releases the resources allocated to the mobile.

3.2. General procedures of the 5G-SA

3.2.1. *Initial random access and beam management procedure*

A subset of the total cell bandwidth is referred to as a Bandwidth Part (BWP). A BWP is a contiguous set of physical resource blocks, selected from a contiguous subset of the common resource blocks for a given numerology (μ) on a given carrier. A UE is configured to be in an active BWP.

Each BWP transmits a burst of synchronization signal blocks (SSB). A burst contains one or more time-spaced synchronization blocks. An SSB contains a primary and secondary synchronization signal to identify the cell (PCI) and a BCCH block that transmits the MIB information of the serving cell.

The burst of SSB is intended to emit a synchronization beam sweeping in different directions. The mobile receives one or more SSB blocks with different received power levels spaced in time.

The maximum number of SSB blocks per burst depends on the carrier frequency:

– 4 SSB blocks maximum for a frequency lower than 3GHz;

– 8 SSB blocks maximum for a lower frequency between 3 GHz and 7.125 GHz;

– 64 SSB blocks maximum for a frequency in the FR2 band.

During the cell search procedure, the mobile searches for the initial BWP partition. The initial BWP partition is a frequency sub-band on which the mobile reads the minimum system information RMSI (Remaining Minimum System Information).

MIB information is carried over the BCCH channel, and SIB1 information is carried over the PDSCH control channel. The SIB1 information allows the mobile to learn about the parameters of random access to the cell.

The physical downlink control channel (PDCCH) carries Downlink Control Information (DCI). The DCI message indicates the downlink resources for PDSCH.

The mobile searches DCI information in a common search space called the Type0-PDCCH search space. According to the k_{ssb} (*ssb-subcarrierOffset*) value transmitted on the MIB message, the mobile knows whether Type0-PDCCH is present or not in the BWP:

– in the case of the FR1 band (Frequency Range 1 ranging from 450 MHz to 7.125 GHz), if $k_{ssb} \leq 23$, then the BWP partitioning contains a Type0-PDCCH on which the mobile reads the SIB1 information;

– in the case of the FR2 band (Frequency Range 2 ranging from 24.250 MHz to 52.6 GHz), if $k_{ssb} \leq 11$, then the BWP partitioning contains a Type0-PDCCH on which the mobile reads the SIB 1 information.

The SIB1 information carries:

– the radio resources of the CORESET control area (COntrol REsource SET) from the PDCCH-ConfigSIB1 message (in addition to the common search area);

– measurement information for the selection or re-selection of cells;

– the configuration of the random access request.

Once synchronized on an SSB, the mobile transmits a random access request to establish a bearer connection.

The procedure is described by the exchange of the following four messages (Figure 3.7).

The MAC entity of the mobile chooses a random preamble from the list corresponding to the serving cell:

1) The mobile sends its random access request by transmitting a preamble. Once the preamble has been sent, the terminal UE listens to the response of the base station between the instant t_1 and $t_2 = t_1 +$ Window_reception (timer T300).

The RA-RNTI identity is calculated as follows:

– $1 + s\_id + 14 * t\_id + 14 * 80 * f\_id + 14 * 80 * 8 * ul\_carrier\_id$;

– s_id is the index of the first OFDM symbol $(1 \leq s\_id \leq 14)$ of the PRACH channel;

– t_id is the index of the first slot of the PRACH channel ($0 \leq t\_id \leq 80$) in the frame;

– f_id is the index of the PRACH physical channel in the frequency domain ($0 \leq f\_id \leq 8$);

– $ul\_carrier\_id$ is the subcarrier index that carries the preamble.

2) When the radio node receives the PRACH physical channel, it calculates the timing advanced (TA) and transmits the MAC RAR (Random Access Response) response to the mobile (message 2), allocating the radio resources to it for the next uplink message (UL Grant) and giving a temporary identity TC-RNTI (Temporary Cell RNTI).

The gNB broadcasts the response on the PDCCH Channel. All of the mobiles received the response, but the CRC of the DCI message is scrambled by the RA-RNTI identifier:

– the mobile retrieves the DCI information from the RA-RNTI identifier;

– the mobile checks whether the preamble RAPID identifier (Random Access Preamble IDentifier) transmitted by the gNB entity corresponds to the preamble initially chosen by the mobile;

– the mobile retrieves the description of its data in the physical PDSCH channel from the DCI information.

In message 3, the mobile sends an RRC *ConnectionRequest* with a UE identifier. This identifier allows the gNB base station to resolve conflicts.

When the gNB entity receives the RRC *SetupRequest* message, it responds to the mobile with the RRC *Setup* (message 4) comprising:

– DCI information (Downlink Control Information) in the physical PDCCH channel;

– the MAC RAR header containing the UE CRI (Contention Resolution Identity) control element. This control element replicates the identity of the RRC Request message, thereby resolving the collision.

The mobile retrieves the DCI message with the TC-RNTI identifier and reads the description of the data in the physical PDSCH channel.

After collision resolution, the temporary TC-RNTI identifier becomes the definitive C-RNTI identifier assigned to the mobile. For others, the msg4 response waits until the timer expires. A new access request will then be renewed within the limits of the requests authorized in the SIB2 message.

The aim of the beam management procedure is to pair a beam emitted by a set of TRxP transmission points with a beam from the mobile. The beam management procedure is carried out in four steps:

– beam sweeping;

– measurement of the power level of the beams;

– determination of the beam;

– the beam measurement report.

In the case of initial access, the choice of beam is based on the reference signals of the SSB block (SSS, PSS and PBCH DMRS).

When the mobile is in the RRC_CONNECTED state, the reference signals CSI-RS (DL) and the SRS (UL) are used to pair the beams.

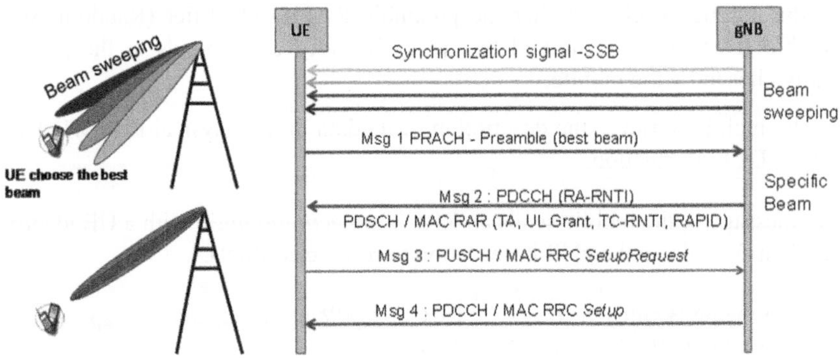

Figure 3.7. *Access radio procedure and beam management. For a color version of this figure, see www.iste.co.uk/launay/5g.zip*

3.2.2. *Establishment of radio connection*

Like the 4G network, following the random access procedure, the mobile requests the establishment of an RRC connection by sending the RRC *SetupRequest.*

The purpose of this request is to inform the base station of the reason for its RRC connection request and to transmit its identifier (ng-5G-TMSI or a random value).

The RRC connection establishment can occur due to:

– an exchange of signaling requests to the core network (MO signaling);

– an emergency call request;

– a non-priority IoT transmission request (delay tolerant);

– a response to a notification (paging: MT access);

– sending an SMS (MO SMS and SMSoIP);

– sending an MDT;

– a request for a telephone call or videoconference from the mobile (MO MMTel Voice, MO MMTel Video).

The base station can accept or refuse the request of the mobile (RRC *SetupMessage* or RRC *RejectMessage*) depending on the importance of the message and the congestion of the radio access or congestion of the core network.

The mobile acknowledges the received message by responding with the RRC *SetupComplete* message.

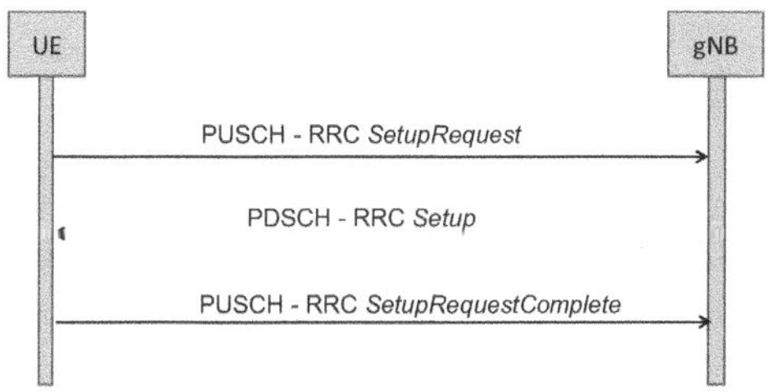

Figure 3.8. *Establishment procedure for the RRC connection*

3.2.3. *Register request*

Excepting emergency calls, the first procedure for the mobile is to register with the 5G core network.

The registration procedure allows the mobile to use 5G network services.

The registration procedure is divided into three steps:

– mutual authentication: identification and authentication;

– access to services;

– mobile registration.

Once registered, the mobile will be able to send or receive its traffic through an E2E bearer. The initial bearer establishment is at the initiative of the mobile and its reestablishment can be implemented by the mobile or by the core network.

3.2.3.1. *Mutual Identification and authentication*

During the RRC connection procedure (Figure 3.8), the mobile sends the RRC *SetupRequest*. This message is used to initiate the Register Procedure by sending the Register Request as a NAS Payload (signaling message).

After receiving the RRC Setup message from the base station, the mobile sends the RRC *SetupComplete* request with the following parameters:

– the message NAS: Registration Request;

– its identifier: SUPI/SUCI (Subscription Permanent identifier/Subscription Concealed identifier) or 5G-GUTI (5G-Global Unique Temporary identifier);

– the expected S-NSSAI (Single Network Slice Selection Assistance Information) network slices;

– its capacity;

– the list of PDU sessions to activate;

– its previous location (TAI).

Both the NAS request and identity of the mobile are transferred from the 5G base station to the core network (AMF: Access and Mobility Management Function).

The AKA (Authentication and Key Agreement) mechanism has been defined to allow the mutual authentication of third parties and distribution of keys when attaching the mobile to the 5G mobile network.

Mutual authentication is based on the AUTN (Authentication Network) and RES (Result) seals generated by the UDM (Unified Data Management) entity and by the mobile. Authentication is based on a RAND sequence and a secret key K_i.

The RAND sequence is generated by the UDM entity and is transmitted to the MME, which then forwards it to the UE. The secret key K_i is stored in the USIM module of the UE's UICC (Universal Integrated Circuit Card) and is provisioned in the UDM entity when the subscription is taken out.

The activation of NAS security is performed by calculating the encryption and integrity key for the exchange of messages between the AMF function and the mobile.

3.2.3.2. *Registration and access to service*

After the activation of NAS security, the AMF function retrieves all of the authorized mobile services (authorized NSSAI) that are provisioned in the UDM. Depending on the location of the user, the AMF function also retrieves the service access policy rules.

The AMF function then proceeds the activation of security on the radio bearer.

The activation of AS security is intended to protect the radio interface:

– with the integrity and ciphering of RRC signaling messages;

– with the integrity of the traffic data (optional);

– by ciphering the traffic data.

When the mobile confirms the activation of the radio bearer, the context of the mobile is registered in the AMF function.

3.2.3.3. *Description of the registration procedure*

The registration procedure is described in Figures 3.9 and 3.10.

1) The registration procedure is triggered by the mobile. After the establishment of the signaling radio bearer, the mobile sends the dedicated NAS message: REGISTRATION REQUEST to the gNB. The message contains the identifier of the mobile (SUPI/SUCI or 5G-TMSI), the previous TAI if it exists, the expected S-NSSAI slices, the capacity of the mobile and the list of PDU sessions if it exists:

– the REGISTRATION REQUEST message is carried by the RRC SetupComplete request on the radio interface NR, and by the NGAP Initial UE Message on the NG-C interface;

– the NGAP Initial UE Message contains user location information, such as the NCGI cell identifier (NR Cell Global Identifier) and the identity of the TAI location area of the mobile.

2) When the AMF function receives the REGISTRATION REQUEST message, it either contacts the UDM entity (if the mobile has transmitted its private SUCI or SUPI identifier) or the AMF function that the mobile was previously registered on (if the mobile has transmitted its 5G identifier ng-GUTI):

– if the mobile was already registered on the network, it sends its 5G-GUTI identifier instead of the SUCI or SUPI. From 5G-GUTI, the gNB entity transmits the registration request to the AMF function that the mobile was previously registered on. If this AMF function cannot be reached, the gNB entity chooses another AMF function. This new AMF function will contact the previous AMF function that the mobile was previously registered on. The goal is to be able to recover the SUCI or SUPI identifier. If the AMF function removed the context of the mobile, then the AMF is no longer able to provide the SUPI or SUCI identity. In this case, the AMF function requests that the mobile provides its private identity IMSI by using the NAS *IdentityRequest* request. This message is carried by the RRC *DLInformationTransfer* request on the radio interface NR, and by the NGAP DL NAS Transport message on the NG-C interface;

– the mobile sends its private identity in the NAS *IdentityResponse* request carried by the RRC *ULInformationTransfer* request on the radio interface NR, and by the NGAP UL NAS *Transport* message on the NG-C interface.

At the end of this procedure, the AMF function contacts UDM to obtain an authentication vector made up of a random number RAND, the authentication response expected from the mobile RES, the authentication token of the AUTN network, a 5G key (ngKSI) and the ABBA parameter deriving from the AMF (Anti-Bidding down Between Architectures) key.

3) The AMF function transmits the *AUTHENTICATION REQUEST* message containing the random number RAND, the seal of the AUTN network and the parameters ngKSI and ABBA to the mobile.

The mobile locally calculates, from its private and secret key K_i (stored in the USIM module of its UICC card) and the random number RAND received, the seal of the AUTN network, which it compares with the value received. If the two values are identical, the 5G core network is authenticated and the mobile calculates, with the additional parameters ngKSI and ABBA, its RES authentication seal and the K_{ASME} key:

– the *AUTHENTICATION REQUEST* message is carried by the RRC *DLInformationTransfer* message on the NR radio interface, and by the NGAP *DOWNLINK NAS TRANSPORT* message on the NG-C interface;

– the *AUTHENTICATION RESPONSE* message is carried by the RRC *ULInformationTransfer* message on the NR radio interface, and by the NGAP *UPLINK NAS TRANSPORT* message on the NG-C interface.

4) The security of the NAS signaling is activated by the AMF function, which sends the *NAS SECURITY MODE COMMAND* message, checked for integrity with the K_{NASint} key. This message contains the algorithms used to derive the K_{SEAF} (SEcurity Anchor Function) key.

The K_{SEAF} key is used to generate the encryption and integrity keys for NAS requests and generate the KgnB key:

– the *NAS SECURITY MODE COMMAND* message is carried by the RRC *DLInformationTransfer* message on the NR radio interface, and by the NGAP *DOWNLINK NAS TRANSPORT* message on the NG-C interface;

– the *NAS SECURITY MODE COMPLETE* message is carried by the RRC *ULInformationTransfer* message on the NR radio interface, and by the NGAP *UPLINK NAS TRANSPORT* message on the NG-C interface.

5) Following the security of the NAS signaling, the 5G core network:

– checks that the mobile is not banned (blacklisted);

– transfers the UE context to the AMF function which registers the mobile;

– updates policy rules at the UE context level;

– establishes the UPF (User Plane Function) transport plan of the PDU sessions to be activated (if they exist).

At the end of these operations, the AMF function transmits the UE context to the NG-RAN access, including:

– the derivation key for securing the K_{gNB} radio link;

– the mobility restriction list;

– the mobile radio capacities;

– the security algorithms used by the mobile;

– the authorized NSSAI indicator;

– the RSFP index (RAT Frequency Selection Priority);

– information concerning PDU sessions (identity of the session, S-NSSAI indicator).

The establishment of the UE context to the radio node is initiated by the AMF function via the NGAP *Initial Context Setup Request.*

6) Using the K_{gNB} security key, the base station secures the radio link via the RRC *SecurityModeCommand* request.

On receiving this message, the mobile also derives the K_{gNB} radio access security key and:

– calculates the integrity key of RRC signaling;

– checks the integrity of the message received;

– derives the encryption key from the radio link;

– starts the procedure for securing the SRB signaling radio support.

The mobile responds with the RRC *SecurityModeComplete* message if the integrity check is successful. At this step, the messages exchanged are protected in integrity but are still not ciphered.

7) The base station responds to the first message transmitted by the mobile (RRC *SetupComplete REGISTRATION REQUEST*) via the RRC *Reconfiguration* request, which carries the *REGISTRATION ACCEPT* message.

The RRC Reconfiguration message also enables us to:

– configure the measurement report at the mobile level;

– add secondary cells in the case of carrier aggregation (SCells);

– configure the DRB (Data Radio Bearer) radio support.

The base station confirms the registration of the UE context to the AMF function with the NGAP Initial Context Setup Response message. This step follows the successful securing of the radio link.

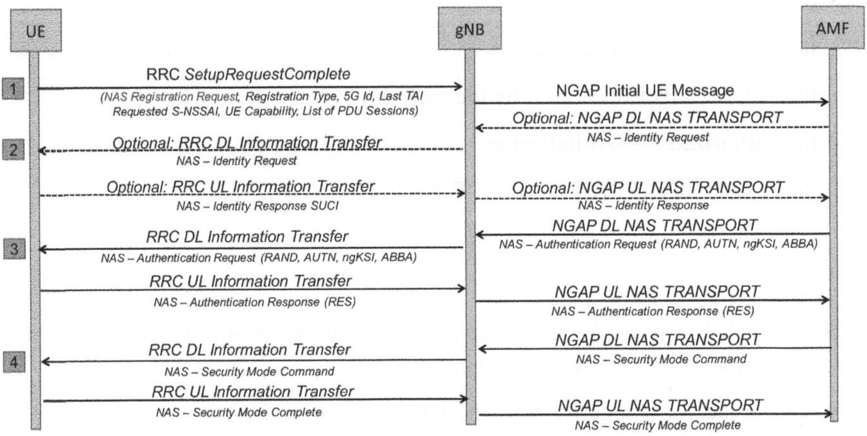

Figure 3.9. *Registration procedure: authentication and the NAS secure mode. For a color version of this figure, see www.iste.co.uk/launay/5g.zip*

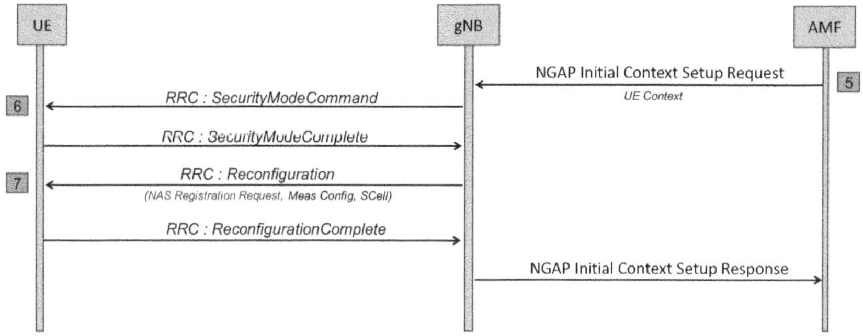

Figure 3.10. *Registration procedure – service access and registration. For a color version of this figure, see www.iste.co.uk/launay/5g.zip*

3.2.3.4. *The AMF function reallocation procedure*

The NG interface management procedure allows NG-C interoperability information to be exchanged between the NG-RAN node and the core network. The network node interrogates the AMF function via the NG *SETUP REQUEST*, in order to find out the S-NSSAI network slices supported by the AMF function on a TA (tracking area).

The AMF informs the NG-RAN node in the NG *SETUP RESPONSE* (Figure 3.11).

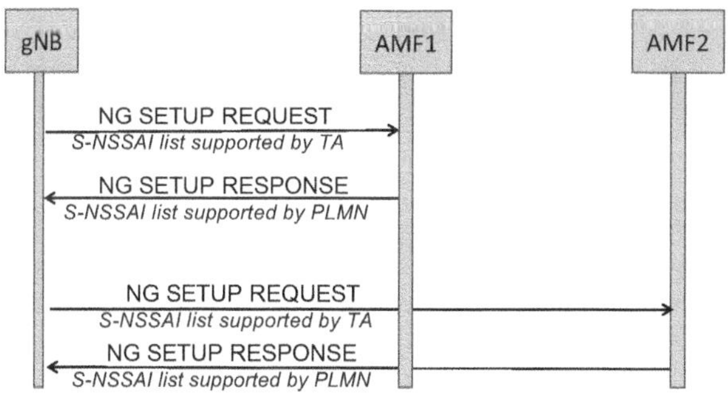

Figure 3.11. *The NG-C interface configuration procedure*

If the AMF function should be updated, to ensure interoperability on the NG-C interface, the AMF function informs the NG-RAN nodes of the corresponding S-NSSAI and TAI evolution (Figure 3.12).

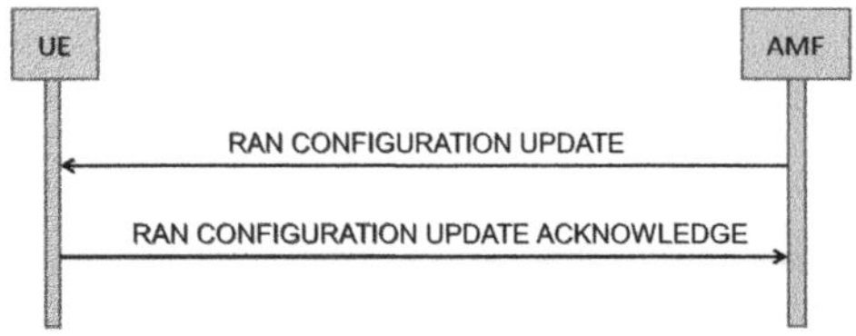

Figure 3.12. *Procedure for updating the AMF function*

The mobile transmits its requested S-NSSAI slices to the gNB in the RRC *ConnectionSetupComplete* message.

The gNB entity selects the AMF function from the NSSAI flag when different AMF are candidates. The gNB entity load balances the source AMF function from among all of the candidates.

However, the selection made by the gNB entity is not necessarily optimal.

The source AMF function which receives the attachment request first checks the S-NSSAI network slices authorized for the mobile, using the UDN database. From S-NSSAI, the AMF queries the NSSF network slice selection function (Network Slice Selection Function) and the NRF (Network Repository Function) discovery function in order to find out the target AMF function offered by the core network.

If the target AMF function for registration is different from the source AMF function, a re-selecting procedure is triggered. The procedure for re-selecting the AMF function is described in Figure 3.13 for option A, and in Figure 3.14 for option B.

In the case of option A, depending on the subscription and local policy information, the source AMF decides to send the initial request to the target AMF via the Namf_Communication_N1MessageNotify message carrying the NG-RAN Reroute Message.

However, since there can only be one end point N2 between the gNB and the AMF for a given UE, the target AMF will update the end point to the gNB via the NG AMF Configuration message.

In the case of option B, depending on the subscription and local policy information, the source AMF decides to send the NGAP *Reroute NAS Message* to the gNB entity so that it formulates a new request for attachment to the target AMF function.

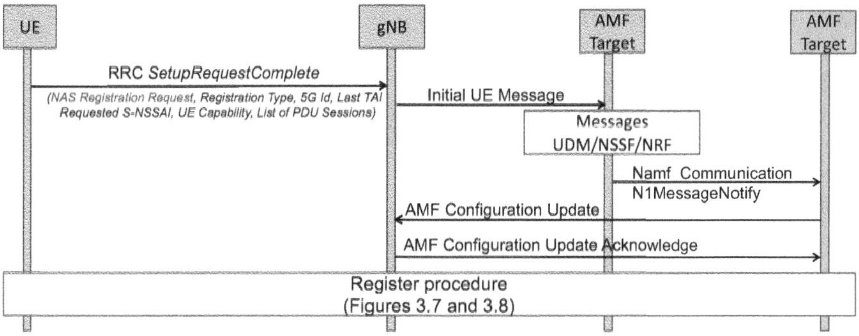

Figure 3.13. *The registration procedure option A: selection and reallocation to a target AMF*

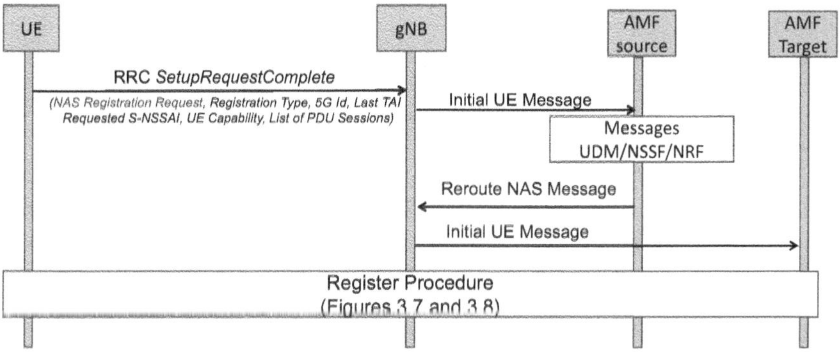

Figure 3.14. *The registration procedure – option B: selection and reallocation of the AMF function by the gNB*

The context of the mobile UE must be transmitted to the target AMF function.

If the UE context is transmitted to the target AMF in the *Reroute NAS Message* carried by the Namf_Communication_N1MessageNotify message, then the target AMF can start the authentication procedure.

If the target AMF does not have context regarding the mobile UE, the source AMF sends the Namf_Communication_UEContextTransfer message to the target AMF to retrieve it.

3.2.4. *The procedure for establishing a PDU session*

During the registration phase, the mobile transmits the NAASI indicator of the desired network in the RRC *SetupRequestComplete* request. The network core determines the authorized network slice, depending on the radio access network.

Once registered, the mobile will be able to establish a PDU (Protocol Data Unit) session for each S-NSSAI network slice. A PDU block is a structured data block consisting of one or more bytes.

A PDU session is defined by three parameters: an S-NSSAI network slice indicator, a Data Network Name (DNN) selection and a transport format type.

The DNN value is the equivalent of the APN (Access Point Name). It allows the core network to choose the SMF function that will manage the IP flows inside this PDU session.

The transport type of a PDU session can be IPv4 or IPv6 Internet connectivity, Ethernet connectivity or unstructured connectivity, such as the encapsulation of data in a point-to-point tunnel.

The core network applies a default QoS for the PDU session. Depending on the operator's policy rules, the core network can also apply specific QoS for a set of flows in the PDU session.

The SMF function will select the function of the UPF transport plane, which will relay the user traffic in the core network with the requested QoS. The application of QoS at the UPF level is achieved by transferring the routing rules defined by the SMF function into the table of the UPF function.

3.2.4.1. *The session establishment procedure*

The session establishment procedure is performed to establish a new PDU session or to reestablish an existing PDU session.

The establishment of a PDU session is:

– initiated by the mobile on the NG-RAN network;

– initiated by the mobile following a handover between either the NG-RAN network and the non-3GPP network, or the 4G mobile network and the 5G mobile network;

– initiated by the network, which notifies the mobile of a request to establish a session.

Establishing a new PDU session can only be initiated by the mobile.

When the network initiates the re-establishment of an existing PDU session, the core network sends paging, informing the mobile of the reason for the re-establishment of the session. The mobile responds with an NAS request to establish a PDU session.

1) The establishment of a PDU session by the mobile is carried out by an NAS PDU *Session Establishment Request* between the mobile and the AMF function, which transfers the request to the SMF function.

The NAS PDU Session Establishment Request message is carried by the RRC *SetupRequestComplete* message on the 5G-NR radio interface and by the NGAP *UEInitialMessage* on the N2 interface.

The mobile transmits:

– the indicator S-NSSAI for the PDU session and the identity of the PDU session if it exists. If the PDU Session ID does not exist, the mobile generates a new PDU Session ID;

– the SSC (Session and Service Continuity) mode of requested service, the DN (Data Network) access name and PCO (Protocol Configuration Options) configuration options to define the address of a DNS server or the IP address of a specific server (P -CSCF, for example);

– the type of session establishment: either an Initial Request for the creation of a new session or a request to re-establish an existing session (Existing PDU Session). It can also be a request to establish a new session for an Emergency Request;

– the capacities of the mobile 5GSM core network concerning:

- the supported transport format (Ethernet),

- QoS management (reflective QoS),

- IPv6 multi-homing sessions,

- ATSSS (Access Traffic Steering, Switching, Splitting) flow routing capabilities for multi-access connectivity services (3GPP and not 3GPP simultaneously),

- the TAI location of the mobile and the type of access network.

2) The radio node transfers the NAS request to the mobile core network via the NGAP *INITIAL UE MESSAGE*.

The AMF function orchestrates the request from the mobile and transmits the order to the SMF function to establish the PDU session according to the QoS criteria relating to the mobile and its location.

The SMF function will inject the routing table at the level of the UPF function (or UPF functions). This table contains the routing rules (tunnel, next hop) associated with the incoming packets (tunnel number and information relating to the PDU format), the QoS to apply and the data volume measurement rules.

In order to establish the tunnel relative to the PDU session, it is necessary to exchange the tunnel identifier TEID and the IP address (between the radio node and the UPF) between the base station and the UPF.

3) The AMF starts the PDU resource session establishment procedure by transmitting the information allowing us to configure the radio node level, the PDU session (tunnel identifier and the IP address of the UPF function) and the resource reservation (scheduling rules) to the NG-RAN node. The information is transmitted in the NGAP *PDU SESSION RESOURCE SETUP REQUEST* message.

4) The radio node transmits the information concerning the establishment of the DRB data radio bearer in the RRC *Reconfiguration* message to the mobile.

5) The mobile acknowledges the receipt of the DRB configuration via the RRC *ReconfigurationComplete* message.

6) The radio node transmits the configuration information of the UPF transport plan (TEID, IP address of the radio node) to complete the routing table to the AMF function. The AMF function transfers this information to the SMF function, which manages the UPF function.

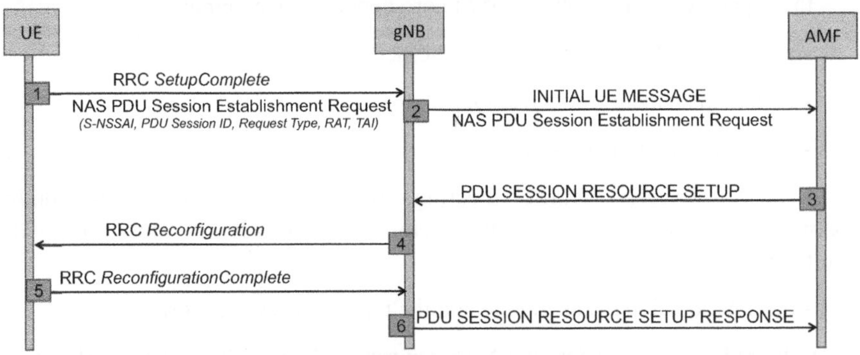

Figure 3.15. *The procedure for establishing a PDU session*

3.2.4.2. *Continuity of the session service*

The continuity of the service of a PDU session is configured during the establishment of the PDU session according to the value of the SSC mode. The continuity of the service makes it possible to guarantee (or not) the conservation of the IP address in the event of mobile mobility, according to one of the following three modes (Figure 3.16):

– SSC mode 1: the core network ensures the continuity of service by keeping the IP address for an IP session;

– SSC mode 2: the core network releases the existing PDU session and therefore the mobile connection to re-establish it on another UPF function. In the case of an IP session, the IP address assigned to the mobile will change;

– SSC mode 3: the core network establishes two PDU sessions simultaneously on two different UPF functions. The mobile temporarily has two IP addresses for an IP session.

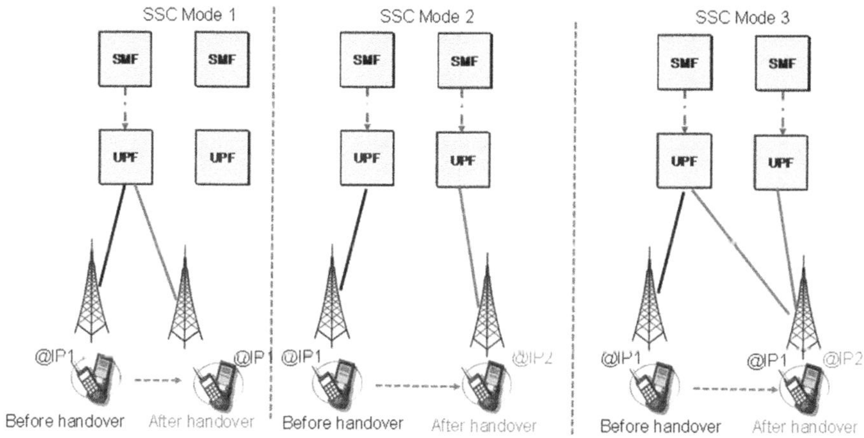

Figure 3.16. *Service continuity. For a color version of this figure, see www.iste.co.uk/launay/5g.zip*

3.3. References

All standards can be downloaded from the ETSI website: https://www.etsi.org/standards.

3GPP TS 23.502
5G - Procedure for the 5G System;
Version 15.3.0 Release 15

3GPP TS 37.340
5G NR - Multi-connectivity; Overall description; Stage-2
Version 15.8.0 Release 15

3GPP TS 38.211
5G NR - Physical Channels and modulation;
Version 15.8.0 Release 15

3GPP TS 38.321
5G, NR - Medium Access Control (MAC) protocol specification;
Version 15.6.0 Release 15

3GPP TS 38.331
5G, NR - Radio Resource Control (RRC); Protocol specification;
Version 15.6.0 Release 15

3GPP TS 38.413
5G, NG-RAN - NG Application Protocol (NGAP);
Version 15.6.0 Release 15

Campos, J. (2017). Keysight: Understanding the 5G NR physical layer [Online]. Available at: https://www.keysight.com/upload/cmc_upload/All/Understanding_the_5G_NR_Physical_Layer.pdf.

4

5G-NR Radio Interface – The Physical Layer

4.1. 5G-NR radio interface

4.1.1. *OFDM waveform*

OFDM modulation consists of modulating orthogonal subcarriers (Figure 4.1).

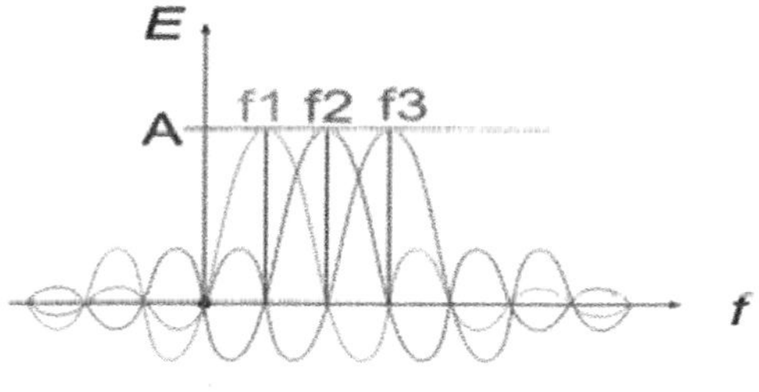

Figure 4.1. *OFDM modulation on three subcarriers*

The two subcarriers $c_1(t)$ and $c_2(t)$ are orthogonal if and only if their scalar product is null: $< c_1(t), c_2(t) >= \frac{1}{T}\int_0^T \cos(2\pi f_1 t)\cos(2\pi f_2 t)dt = 0$.

Orthogonality conditions impose that $T = \frac{1}{\Delta f}$, where Δf is the subcarrier spacing and T represents the duration of a symbol.

For a color version of all the figures in this chapter, see www.iste.co.uk/launay/5g.zip.

Thanks to orthogonality, OFDM modulation offers good spectral efficiency and provides resilience to the frequency selective fading of the propagation channel.

The OFDM signal is used on the downlink as it allows multiple access capacities according to the OFDMA (Orthogonal Frequency Division Multiple Access) technique. The zero-padding technique makes it possible to reduce the PAPR (Power Average Power Ratio) and the power rejection on the neighboring bands. The PAPR represents the ratio of the maximum power of the OFDM signal to the average power in a bandwidth.

The uplink transmission uses either the OFDM signal or the DFT-S-OFDM (Discrete Fourier Transform Spread OFDM) signal. The DFT-S-OFDM signal is derived from OFDM modulation using transform precoding and is applied over a wider frequency band than the OFDM signal. The multiplexing method used in uplink is either OFDMA or SC-FDMA (Single Carrier – FDMA).

The advantage of the DFT-S-OFDM signal is the reduction in power consumption (by reducing the PAPR). On the other hand, the DFT-S-OFDM signal is less spectrally efficient than the OFDM signal. In addition, for modulation with reduced bandwidth, the OFDM signal has been proven to consume less energy than the DFT-S-OFDM signal.

OFDMA and SC-FDMA access multiplexing allow time–frequency resources to be distributed among each user.

Spatial Division Multiple Access (SDMA) multiplexing, also known as the Multi-User MIMO (MU-MIMO) transmission mode, allows simultaneous transmission on the same frequency bands, to multiple users. SDMA multiplexing requires the use of multiple antennas.

From Release R.16, 5G-NR offers non-orthogonal access multiplexing (NOMA), for which the time–frequency resources of the same antenna are used simultaneously by several users. With NOMA, the receiver sets up an MUD (Multi-User Detection) architecture in order to remove interference from other users via the SIC (Successive Interference Cancellation) algorithm. After the SIC algorithm, each user can recover their own signal.

4.1.2. *Frequency bands and multiplexing methods*

Two frequency ranges (FR) are defined:

– the FR1 frequency range covers the frequency bands between 450 MHz and 7.125 GHz (according to Release R.16). The bandwidth of the radio channel is between 5 MHz and 100 MHz (Table 4.1);

– the FR2 frequency range covers the frequency bands between 22.450 GHz and 52.600 GHz (according to R.16 release). The maximal bandwidth of the radio channel is between 50 MHz and 400 MHz (Table 4.2).

5G-NR – FR1 operational bands			
Band	**Frequency [MHz]**	**Bandwidth [MHz]**	**Duplexing method**
n77	3300–4200	10–100	TDD
n78	3300–3800	10–100	TDD
n79	4400–5000	40–100	TDD
n80	1710–1785	5–30	SUL
n81	880–915	5–20	SUL
n82	832–862	5–20	SUL
n83	703–748	5–20	SUL
n84	1920–1980	5–20	SUL
n86	1710–1780	5–40	SUL
n90	2496–2690	10–100	TDD

Table 4.1. *Frequency band FR1 5G-NR*

5G-NR – FR2 operational bands			
Band	**Frequency [GHz]**	**Bandwidth [MHz]**	**Duplexing method**
n257	26.5–29,5	50–400	TDD
n258	24.25–27.5	50–400	TDD
n259	39.5–43.5	50–400	TDD
n260	37–40	50–400	TDD
n261	27.5–28.35	50–400	TDD

Table 4.2. *Frequency band FR2 5G-NR*

Duplexing defines how to share the communication channel between the DL (downlink) and the UL (Uplink) transmission.

There are three main duplexing modes (Figure 4.2) and a hybrid mode:

– FDD (Frequency Division Duplexing);

– TDD (Time Division Duplexing);

– FD (Full Duplex) spatial duplexing;

– SUL (Supplementary Uplink) mode.

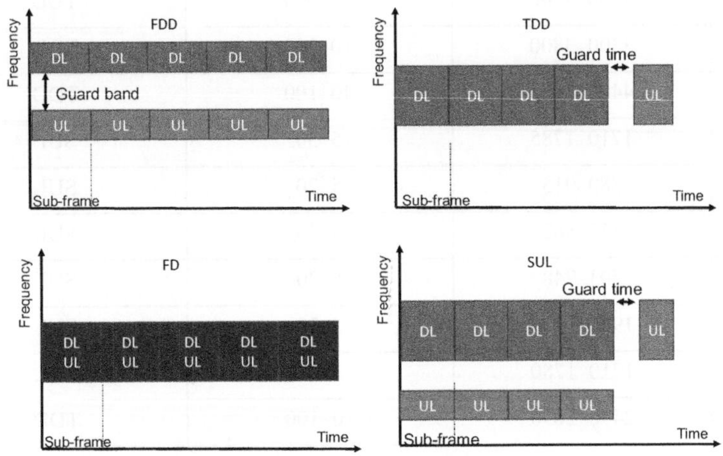

Figure 4.2. *5G-NR multiplexing techniques*

FDD multiplexing requires two frequency paired bands, one band for the downlink transmission and another for the uplink transmission. The downlink and uplink bands are separated by a frequency guard band. The frequency band used by the mobile for the uplink channel is controlled by the base station, and the control information is transmitted to the mobile on the paired downlink band.

For TDD multiplexing, the two directions of transmission share the same frequency band, temporally. TDD multiplexing can increase the throughput of the downlink at the expense of the uplink by allocating more time for the downlink transmission. TDD is better than FDD in exploiting the massive-MIMO technique.

For FD multiplexing, the two directions of transmission occur simultaneously in the same band, exploiting the spatial diversity of the radiating elements: one part of the radiating elements is configured for transmission and the other part for reception.

The 5G-NR radio interface offers the SUL transmission mode: an additional frequency band is introduced, just for the uplink. Unlike FDD multiplexing, the SUL uplink is paired with any downlink band. This mode improves the coverage area for the uplink and reduces latency. In this mode, the mobile cannot transmit on the two uplink bands simultaneously (TDD and SUL or FDD and SUL).

The three transmission modes TDD, FDD and SUL are supported in the frequency range FR1. The FR2 frequency range only supports the TDD mode.

The bandwidth capacity of the mobile is not necessarily as wide as the base station's radio bandwidth. The bandwidth can be adjusted at any time.

4.1.3. *NR frame structure*

OFDM modulation has been implemented on the LTE (Long-Term Evolution) radio interface. In the case of LTE, the spacing Δf between the subcarriers (SCS) has a single value equal to 15 kHz.

The symbols at the input of the OFDM modulator are distributed by a serial/parallel converter and transmitted to the DSP (Digital Signal Processor), which performs the operation of the IFFT (Inverse Fast Fourier Transform).

For the LTE interface, the size of the IFFT inverse Fourier transform is 2048 subcarriers, but only a maximum of 1200 subcarriers are modulated (except for the case of NB-IoT in the guard band). The duration of an OFDM symbol is inversely proportional to the spacing between subcarriers; the choice of 15 kHz results in a sample rate of 30.72 MHz. This value guarantees clock compatibility with 3G (3.84 MHz) and 2G (200 kHz).

For the NR interface, the spacing between the subcarriers Δf takes several values depending on the parameter μ, according to the formula $\Delta f = 2^{\mu} \times 15$ [kHz].

The spacing between the subcarriers with the frequency bandwidth defines the IFFT size (Table 4.3). The size of the 5G-NR IFFT is variable between 512, 1024, 2048 or 4096, with a limitation to 3300 subcarriers:

– μ = {0, 1}: these values are reserved for the FR1 frequency range;

– μ = {2}: this value is used for the FR1 and FR2 frequency ranges;

– μ = {3, 4}: these values are reserved for the FR2 frequency range.

5G-NR – Numerology					
SCS\BW	**20 MHz**	**50 MHz**	**100 MHz**	**200 MHZ**	**400 MHZ**
15 kHz	2048 IFFT	4096 IFFT			
30 kHz	1024 IFFT	2048 IFFT	4096 IFFT		
60 kHz	512 IFFT	1024 IFFT	2048 IFFT	4096 IFFT	
120 kHz		512 IFFT	1024 IFFT	2048 IFFT	4096 IFFT

Table 4.3. *FFT length according to the 5G bandwidth and SCS*

4.1.4. *NR frame structure in the time domain*

The downlink and uplink transmissions are organized into frames of 10 ms duration, each divided into 10 1 ms subframes. Each frame is also divided into two half-frames of size equal to five subframes:

– half-frame 0 is made up of subframes 0–4;

– half-frame 1 is made up of subframes 5–9.

For the LTE interface, the subframe is made up of two time slots. For the NR interface, the number of time slots (slots) constituting the subframe depends on the spacing between the subcarriers (Table 4.4).

For the LTE interface, the value of the Interval Transmission Time (ITT) corresponds to the duration of the subframe and has a fixed value of 1 ms. For the NR interface, the slot duration has a value that depends on the frequency spacing between the subcarriers (Table 4.4).

Subcarrier subspacing	Number of *slots* per subframe	Number of *slots* per frame	TTI
15 kHz	1	10	1 ms
30 kHz	2	20	0.5 ms
60 kHz	4	40	0.25 ms
120 kHz	8	80	0.125 ms
240 kHz	16	160	0.0625 ms

Table 4.4. *Structure of the NR frame in the time domain*

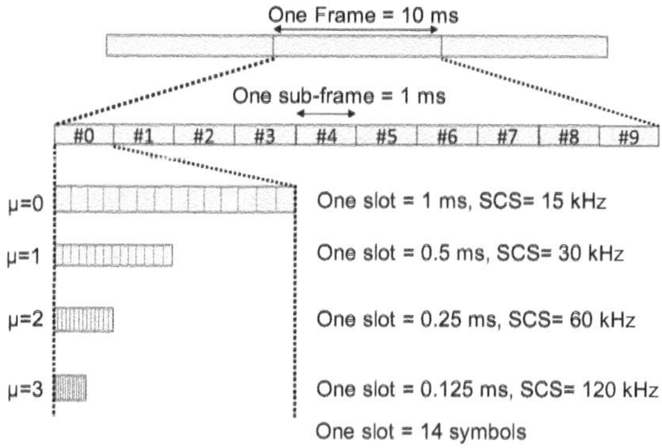

Figure 4.3. *Time structure of the 5G-NR frame*

To avoid interference between two symbols, the OFDM modulator inserts a symbol replica at the start of each symbol. This replica constitutes the Cyclic Prefix (CP) and occupies a fixed percentage in each time slot.

For the 5G-NR access interface, there are two types of prefix (Table 4.5):

– the normal cyclic prefix, which applies to all numerologies;

– the extended cyclic prefix, which only applies for the numerology μ = 2. The extended cyclic prefix improves the robustness of communication in a time dispersive channel, but reduces spectral efficiency.

For the LTE interface, a slot lasts 0.5 ms and includes seven OFDM symbols for the normal cyclic prefix and six OFDM symbols for the extended cyclic prefix.

For the NR interface, the slot includes 14 OFDM symbols for the normal cyclic prefix and 12 OFDM symbols for the extended cyclic prefix.

μ	$\Delta f = 2^{\mu} \cdot 15\,[\text{kHz}]$	**Cyclic prefix**
0	15	Normal
1	30	Normal
2	60	Normal, Extended
3	120	Normal
4	240	Normal

Table 4.5. *Prefix cyclic per numerology*

4.2. TDD mode configurations

For the TDD mode, the elementary resource allocated to the downlink or uplink is different between LTE and NR interfaces:

– for the LTE interface, the elementary resource corresponds to a subframe and seven configurations are defined at the subframe level;

– for the NR interface, the elementary resource corresponds to an OFDM symbol and several configurations (patterns) are defined;

– for the NR interface, the number of symbols per subframe depends on the numerology (Table 4.4).

The 5G-NR TDD frame provides flexibility in the direction of radio resources per slot: OFDM symbols in a slot can be up (uplink), down (downlink) or flexible. A flexible symbol corresponds to an ascending or descending direction. The format of a slot represents the directivity of the symbols in the slot.

The format configuration of a slot can be defined semi-statically or dynamically.

The dynamic configuration corresponds to the sequencing of the transmission of the mobile based on the control signal DCI (Downlink Control Information). It is also possible to configure an SFI (Slot Format Indicator) slot format for a group of mobiles.

The semi-static configuration is transmitted to the mobile by an RRC control message through two Information Elements (IE):

– IE tdd-UL-DL-ConfigurationCommon is transmitted in the SIB1 (System Information Block) message and defines the cell-specific configuration model;

– IE tdd-UL-DL-ConfigurationDedicated is transmitted to the mobile when configuring the radio channel and defines the mobile-specific configuration template.

If the symbol direction of a slot is not configured, the mobile considers that the symbol transmission direction is flexible (either up or down).

4.2.1. *Static configuration per cell*

The cell-specific configuration is defined by one or two profiles (patterns) transmitted to the mobile via an RRC message.

The RRC message is either broadcasted in the cell by the SIB1 message or transmitted to the mobile during a dedicated RRC message.

The first profile (pattern 1) is mandatory, the second (pattern 2) is optional.

In the time domain, the tdd-UL-DLConfigurationCommon configuration is characterized by:

– the period P of the TDD UL/DL duplexing scheme (0.5 ms, 0.625 ms, 1 ms, 1.25 ms, 2.5 ms, 5 ms, 10 ms);

– the number of DL *dslots* slots (between 0 and 320);

– the number of *dsym* symbols in DL (between 0 and 13);

– the number of UL *uslot* slots (between 0 and 320);

– the number of *usym* symbols in UL (between 0 and 13).

Over one period P, the frame is configured with the following information (Figure 4.4):

– number of symbols in DL: *dslots* slots in DL, followed by *dsym* symbols in DL;

– number of symbols in UL: *usym* symbols in UL, followed by *uslots* slots in UL.

The total number of slots (DL and UL) of the profile depends on the numerology and the period of the multiplexing scheme according to the formula $P*2^{\mu}$, which must be an integer value (thus the value of 0.625 ms is only valid for the numerology $\mu = 3$).

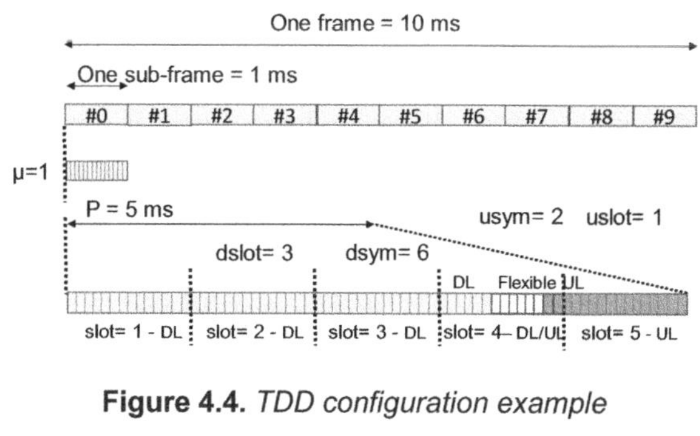

Figure 4.4. *TDD configuration example*

If the two profiles (patterns 1 and 2) are configured, then the mobile configures the direction of the symbols on the first period according to pattern 1 and the second period according to pattern 2.

4.2.2. *Specific TDD configuration*

The base station imposes the cell configuration of the slot format, according to the direction of uplink, downlink or flexible transmission.

In the case where a symbol is configured as flexible, it can be uplink or downlink.

The configuration profile *tdd-UL-DL-ConfigurationDedicated* allows us to define the configuration of the direction of transmission for each symbol.

4.2.3. *The dynamic configuration of the transmission for a group of mobiles*

When the mobile is in the connected state, it is scheduled by the base station through DCI information control messages.

The slot configuration for a group of mobiles is transmitted by an SFI indicator (Slot Format Indicator) carried by the DCI 2.0 message.

The Slot Format Indicator is a pointer to a configuration table that contains 256 values (see Slot formats for normal cyclic prefix – 3GPP TS 38.213 V16.1.0, Table 11.1.1-1). At the level of specification Release 16 (V2.0 – July 2020), only 55 combinations are standardized.

The value SFI = 255 informs the mobile that the configuration of the slot is determined by the static message.

The dynamic slot configuration is reserved for isolated base stations because this configuration makes it possible to allocate radio resources in a very flexible way, via an SFI for a mobile, or for a group of mobiles (e.g. femtocell). The SFI indicator can also indicate the direction of another cell's slot (cross-carrier).

4.3. Physical resource

4.3.1. *Resource grid*

A Resource Element (RE) is the smallest unit that can be assigned to a signal (Figure 4.5). The RE corresponds to an OFDM symbol in the time domain and a subcarrier in the frequency domain. It is thus identified by the pair (k, l), with k representing the index of the subcarrier and l representing the index of the OFDM symbol in the time domain, with respect to a relative reference point.

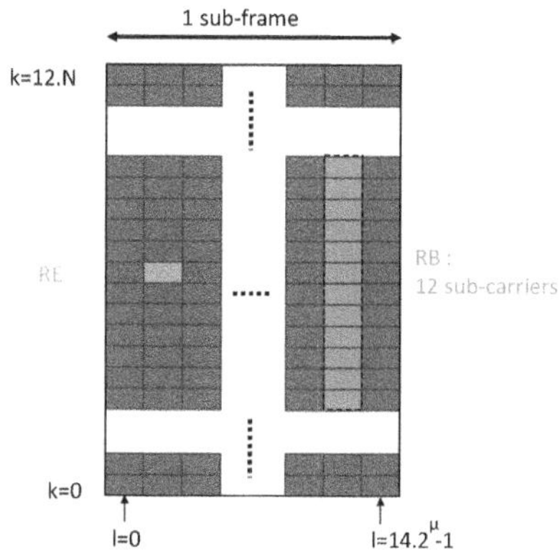

Figure 4.5. *Resource grid*

The RB (Resource Block) corresponds to an allocation of 12 contiguous subcarriers (Figure 4.5). Unlike 4G, the RB 5G resource block corresponds to a frequency allocation.

The resource grid is an allocation of tempo-frequency resources intended for an antenna port. It consists of a set of symbols per subframe (see Table 4.6) in the time space, and a set of contiguous subcarriers in the frequency domain ($N_{grid}^{size,\mu}.N_{sc}^{RB}$).

The resource grid consists of a maximum of 3,300 subcarriers and is transmitted on each direction of transmission and on each antenna port.

Numerology	**Number of symbols in a *slot***	**Number of *slots* per subframe**	**Number of symbols per subframe**
0	14	1	14
1	14	2	28
2	14	4	56
3	14	8	112
4	14	16	224

Table 4.6. *The structure of the resource grid in the temporal domain*

The starting position of the resource grid $N_{grid}^{start,\mu}$ and the spectral width $N_{grid}^{size,\mu}$ are transmitted by the RRC message in the case of 5G-NSA, or via the SIB1 message in the case of 5G-SA.

4.3.2. *Resource bloc and bandwidth part*

Two types of resource block are defined:

– a physical resource block (PRB);

– a common resource block (CRB).

As the resource grid is characterized by its numerology and the number of contiguous PRB resource blocks, it is necessary to define the position of the resource grid $N_{grid}^{start,\mu}$ from the number of the CRB.

The first resource block of the 5G-NR band is located in the frequency domain by a reference point, named point A. The reference point is identified as being the common resource block 0.

The location of point A is transmitted to the mobile by either:

– an offset between the synchronization signal and the start of the 5G band by the offset value *OffsetToPointA*;

– or an absolute reference *AbsoluteFrequencyPointA*, indicating the frequency number ARFCN (Absolute Radio Frequency Channel Number) separating the synchronization signal to the reference point A.

In the case of 5G-NSA, only the absolute reference is transmitted to the mobile in an RRC *ConnectionReconfiguration* message.

In the case of 5G-SA, the mobile retrieves this value from the information message SIB1.

The reference point (Point A) is the starting point for the numbering of the common resource blocks. The position of the CRBs is defined from this point according to their numerology. For a given numerology μ, if k is the number of subcarriers separating a resource block from point A, then the number of the common resource block is defined by:

$$n_{CRB}^{\mu} = \left\lfloor \frac{k}{N_{sc}^{RB}} \right\rfloor$$

The BWP is a collection of contiguous block resources in a resource grid. An ith BWP band partitioning is defined by its numerology and by its bandwidth. The bandwidth corresponds to the number of block resources $N_{BWP,i}^{size}$.

The position of the physical resource blocks (PRB) inside the BWP varies from 0 at the start of the partition to $N_{BWP,i}^{size} - 1$.

A mobile can be configured to support, at most, four BWP partitions per carrier, but only one partition is active in the uplink and downlink directions (the limitation of four BWP concerns the specifications of Releases R.15 and R.16, at least).

The BWP makes it possible to adapt the 5G band to the capacities of the mobile. The mobile can only receive traffic data or control information inside the active BWP partition.

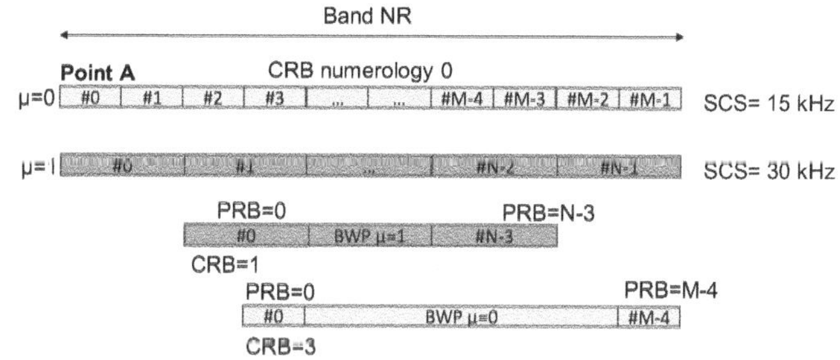

Figure 4.6. *Common resource block and physical resource block*

4.4. Physical channels and physical signals

4.4.1. *Physical signals and reference signals*

The physical PSS (Primary Synchronization Signal) ensures the synchronization of the OFDM signal and the time synchronization at the half-frame level.

The physical SSS (Secondary Synchronization Signal) provides time synchronization at the frame level.

The physical CSI-RS (Channel State Information RS) is a reference signal transmitted by the radio node gNB. The physical signal CSI-RS is specific to UE. It aims for UE to perform a coherent demodulation of the received signal. Channel equalization is based on the calculation of the transfer function from CSI-RS.

The physical SRS (Sound Reference Signal) is a mobile-specific reference signal, allowing the radio node gNB to perform a coherent demodulation of the received signal based on the calculation of the radio channel transfer function.

The physical DM-RS (Demodulation RS) is a mobile-specific signal used to perform a coherent demodulation of the received signal and for beamforming. The DM-RS physical signal is associated with a physical channel.

The physical PT-RS (Phase Tracking Reference Signal) is a reference signal to compensate the frequency difference between the transmitter and receiver and suppress phase noise.

The physical PRS (Positioning Reference Signal) is used by the mobile to determine its location from the OTDOA (Observed Time Difference of Arrival) mechanism.

The physical RIM (Remote Interference Management) signal is used to provide information about interference experienced by other gNB radio nodes, in order to control the DL/UL interference.

4.4.2. *Physical channels*

4.4.2.1. *Downlink direction*

The physical PBCH (Physical Broadcast Channel) transmits the BCH transport channel containing system information corresponding to the MIB (Master Information Block) message. The PBCH channel is time multiplexed with the PSS and SSS channels to form the SSB (Synchronization Signal Block).

The physical PDCCH (Physical Downlink Control Channel) transmits the DCI control information, which allows us to define:

– the allocation of resources, as well as the modulation and coding scheme, for the data contained in the PDSCH (Physical Downlink Shared Channel) and PUSCH (Physical Uplink Shared Channel) physical channels;

– the transmitting power of the PUCCH (Physical Uplink Control Channel) and PUSCH physical channels.

The PDSCH physical channel transmits the DL-SCH (Downlink Shared Channel) and PCH (Paging Channel) transport channels.

The Physical Multicast Channel (PMCH) transmits the MCH transport channel for MBSFN transmissions.

4.4.2.2. *Uplink direction*

The PRACH (Physical Random Access Channel) contains a preamble used by the mobile to ask for the establishment of a radio bearer. The random access procedure is the first step in connecting the mobile to the gNB radio node (see Chapter 3).

The PUCCH (Physical Uplink Control Channel) allows the UE terminal to send UCI (Uplink Control Information) to the gNB radio node, such as:

– HI information (HARQ Indicator) concerning the positive ACK or negative NACK acknowledgment in response to data received on the PDSCH physical channel;

– the ratio of the CSI (Channel State Information) signal used by the gNB to define the transmission mode, as well as the coding and modulation scheme;

– the SR (Scheduling Request) to obtain tempo-frequency resources on the PUSCH physical channel.

The PUCCH and PUSCH physical channels can be multiplexed in the FDM (Frequency Division Multiplex) or TDM (Time Division Multiplex) time domain.

4.5. Downlink transmission

4.5.1. *Synchronization signal*

The SSB block consists of multiplexing physical PSS/SSS synchronization signals with the physical broadcast channel PBCH.

The PSS/SSS signals are mapped to 127 contiguous subcarriers and two symbols.

The BCCH channel uses 240 contiguous subcarriers and two symbols.

The SSB block is formed by multiplexing the SSS/PSS signals with the PBCH channel. The SSB block is mapped to 240 subcarriers and four symbols (Figure 4.8). One or more SSB blocks are transmitted in an SSB burst confined in a half-frame.

An SSB block is periodically transmitted with a configurable period between 5 ms and 160 ms.

The configuration of the SSB burst corresponds to the temporal position of the blocks in a half-frame and to the periodicity of the block.

The configuration of the SSB burst is transmitted to the mobile in the SIB1 message for 5G-SA and in the RRC *Reconfiguration* message for 5G-NSA.

4.5.1.1. *Physical PSS signal*

The physical PSS signal performs the following functions:

– frequency synchronization;

– time synchronization at the level of the OFDM symbol, the time interval (slot) and the half-frame (periodicity 5 ms);

– the determination of the value $N_{ID}^{(2)}$.

The physical PSS is generated from a pseudo-random sequence of 127 bits (Figure 4.7).

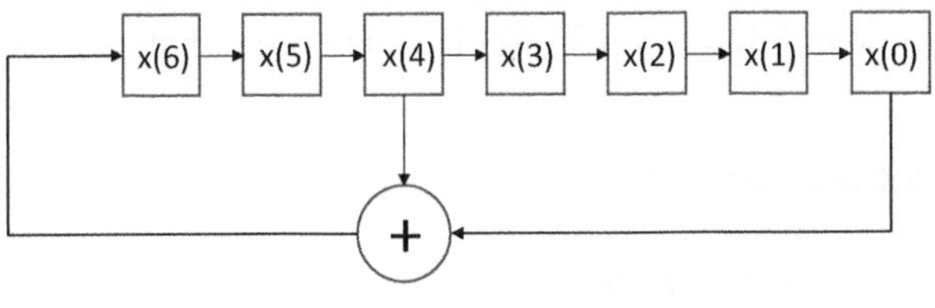

Figure 4.7. *PSS signal generation*

Register initialization: $\left[x(6)\,x(5)\,x(4)\,x(3)\,x(2)\,x(1)\,x(0)\right] = [1110110]$

At each clock rising edge, the bit x(6) is generated by a right shift of the register data. The new value x(6) is calculated by a logical exclusive, or from the values 0 and 4 of the register.

The PSS is an M-sequence, which is characterized by the following equation:

$$x(i+7) = [x(i+4) + x(i)] \bmod 2 \ (0 \le i \le 119).$$

The generated PSS sequence is written as $d(n) = 1 - 2x(m)$, with $m = (n + 43N_{ID}^{(2)}) \bmod 127$.

$N_{ID}^{(2)}$ represents the number of the cell in the group and can take integer values between 0 and 2.

4.5.1.2. *Physical SSS signal*

The physical SSS signal performs the following functions:

– determination of the Cyclic Prefix (CP), normal or extended;

– determination of the number value $N_{ID}^{(1)}$.

N_{ID}^{cell} is the physical identity of the cell PCI (*Physical Layer Cell Identity*). N_{ID}^{cell} can take 1008 values and is calculated by: $N_{ID}^{cell} = 3N_{ID}^{(1)} + N_{ID}^{(2)}$.

The SSS signal corresponds to a Gold sequence of 127 bits among 336 possible sequences. $N_{ID}^{(1)}$ represents the group number (0–335).

The physical SSS signal is generated by the Gold sequence, resulting from the two M-sequences, x_0 and x_1, according to:

$$[x_{0/1}(6)\,x_{0/1}(5)\,x_{0/1}(4)\,x_{0/1}(3)\,x_{0/1}(2)\,x_{0/1}(1)\,x_{0/1}(0)] = [0000001]$$

$$x_{0/1}(i+7) = [x_{0/1}(i+4) + x_{0/1}(i)] \bmod 2.$$

The SSS sequence is generated by:

$$d(n) = \left[1 - 2x_0\left((n + m_0) \bmod 127\right)\right]\left[1 - 2x_1\left((n + m_1) \bmod 127\right)\right]$$

with $m_0 = 15\left\lfloor \frac{N_{ID}^{(1)}}{112} \right\rfloor + 5.N_{ID}^{(2)}$ and $m_1 = .N_{ID}^{(1)} \bmod 112$.

4.5.1.3. *Physical PBCH channel*

The physical PBCH channel transmits the BCH (Broadcast Channel) transport channel, which contains the system information corresponding to the MIB message.

The MIB message is mixed with a scrambling sequence from the Gold sequence and is initialized with the cell identity number:

$$c(n) = \left(x_1(n+N_c) + x_2(n+N_c)\right) \bmod 2$$
$$x_1(n+31) = \left(x_1(n+3) + x_1(n)\right) \bmod 2$$
$$x_2(n+31) = \left(x_2(n+3) + x_2(n+2) + x_2(n+1) + x_2(n)\right) \bmod 2$$

with the vectors initialized as follows:

$$x_1(0,1,..,30) = [1,0,....,0] \text{ and } x_2(0,1,..,30) = C_{init} = N_{ID}^{cell}, \text{ cell identity.}$$

4.5.1.4. *The mapping of SSB*

The SSB block consists of 4 OFDM symbols in the time domain and 240 subcarriers in the frequency domain.

The SSB block is transmitted over antenna port 4000.

The PSS signal is located on the symbol 0 and is spread over 127 subcarriers.

The SSS signal is located on symbol 2 and is spread over 127 subcarriers.

The PBCH signal is mapped on 240 subcarriers at symbol 1 and symbol 3 and split on two frequency blocks at symbol 2 (Figure 4.8).

To help the receiver to demodulate the PBCH signal, a DM-RS signal is associated with the PBCH. The DM-RS signal is transmitted on symbols 1, 2 and 3. In the frequency domain, CSI-RS is transmitted as a comb spaced by three subcarriers, as shown in Table 4.7. The comb helps to take frequency dispersion into account.

The mobile UE performs power measurement on the SSB block (SS-RSRP-Synchronization Signal Reference Received Power) and estimates the quality of the SS-RSRQ (SS Reference Received Quality) radio link from the synchronization signal. The SS-RSRP measurement is averaged over the SS/PBCH Block Measurement Time Configuration (SMTC). The SS-RSRQ measurement is the ratio between the measured power SS-RSRP and the average power received by the mobile NR-RSSI (NR-Received Signal Strength Indicator).

Here, v is the frequency shift (identity number of the modulo 4 cell).

Physical channel or signal	**Number of OFDM symbol**	**Number of the subcarrier**
PSS	0	56,57,…,182
SSS	2	56,57,…,182
Zero-padding	0	0,1,…,55, 183, 184, …239
	2	48,49,..,55, 183, 184, …191
PBCH	1,3	0,1,..,239
	2	0,1,…57 192,193,…239
DM-RS associated to PBCH	1,3	0+v, 4+v,8+v,236+v
	2	0+v, 4+v,…,44+v 192+v, 196+v,…,236+v

Table 4.7. *SSB mapping*

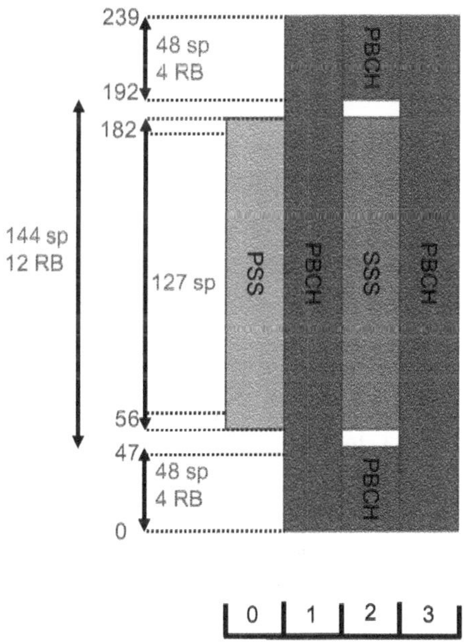

Figure 4.8. *SSB block*

4.5.2. *Reference signals*

4.5.2.1. *CSI-RS signal*

The physical CSI-RS (Channel State Information Reference Signal) signal allows the mobile to measure the received signal (CSI-RSRP) and measure the interference level of neighboring cells.

The physical reference signal is either transmitted with a non-zero power to allow the receiver to determine the level of the received signal NZP-CSI-RS (Non Zero Power CSI-RS), or not transmitted ZP-CSI-RS to allow the receiver to measure the power level of neighboring base stations.

The physical CSI-RS was introduced in Release R.10 to probe the propagation channel on the downlink. In 5G, the CSI-RS are specific to each UE. When switching to the connected mode (RRC Connected), the UE terminal receives the configuration of the CSI-RS. The CSI-RS signal configuration allows the probing of up to 32 antenna ports. Even though the 32 CSI-RS are mobile-specific, multiple terminals can measure the common CSI-RS.

An antenna port is a concept of logical antenna perceived by the mobile and corresponds to one or more physical antennas.

The NZP-CSI-RS signal is a 31-bit pseudo-random Gold sequence which is transmitted to an antenna port. The CSI-RS is generated by the following formula:

$$r(n)=\frac{1}{\sqrt{2}}(1-2.c(2n))+j\frac{1}{\sqrt{2}}(1-2.c(2n+1))$$, c(n) is initialized with:

$$c_{init}=\left(2^{10}(14.n^{\mu}_{s,f}+l+1)(2n_{ID}+1)+n_{ID}\right)\bmod 2^{31}$$

$n^{\mu}_{s,f}$ is the number of the slot in the frame, l is the OFDM symbol position in the slot and n_{ID} is a parameter given by the RRC layer.

Like PBCH, c(n) is a Gold sequence constructed with the following two M-sequences:

$$c(n)=\left(x_1(n+N_c)+x_2(n+N_c)\right)\bmod 2$$

$$x_1(n+31)=\left(x_1(n+3)+x_1(n)\right)\bmod 2$$

$$x_2(n+31)=\left(x_2(n+3)+x_2(n+2)+x_2(n+1)+x_2(n)\right)\bmod 2$$

The CSI-RS occupies, at most, one single resource element (RE) per slot, per RB, and per antenna port at position (k, l), except for the antenna port 3000, for which three resource elements can be used.

The density of the reference signal can be reduced by only transmitting CSI-RS per RB, every two slots. Thus, the density of the reference signal per antenna port, per RB and on a slot is 0.5, 1 (or 3 for only the first antenna port).

In order to measure propagation channels over multiple antenna ports simultaneously, resource elements are reserved for each antenna port, with a limitation to 32 antenna ports per mobile. The antenna ports carrying the CSI-RS signals are numbered from 3000 to 3031.

The mobile should be able to separate the CSI-RS specific to each antenna port.

Multiplexing order	Index number	Code W_f	Code W_f
CDM2	0	$[+1 \ \ +1]$	1
	1	$[+1 \ \ -1]$	1
CDM4	0	$[+1 \ \ +1]$	$[+1 \ \ +1]$
	1	$[+1 \ \ -1]$	$[+1 \ \ +1]$
	2	$[+1 \ \ +1]$	$[+1 \ \ -1]$
	3	$[+1 \ \ -1]$	$[+1 \ \ -1]$
CDM8	0	$[+1 \ \ +1]$	$[+1 \ \ +1 \ \ +1 \ \ +1]$
	1	$[+1 \ \ -1]$	$[+1 \ \ +1 \ \ +1 \ \ +1]$
	2	$[+1 \ \ +1]$	$[+1 \ \ -1 \ \ +1 \ \ -1]$
	3	$[+1 \ \ -1]$	$[+1 \ \ -1 \ \ +1 \ \ -1]$
	4	$[+1 \ \ +1]$	$[+1 \ \ +1 \ \ -1 \ \ -1]$
	5	$[+1 \ \ -1]$	$[+1 \ \ +1 \ \ -1 \ \ -1]$
	6	$[+1 \ \ +1]$	$[+1 \ \ -1 \ \ -1 \ \ +1]$
	7	$[+1 \ \ -1]$	$[+1 \ \ -1 \ \ -1 \ \ +1]$

Table 4.8. *CDM multiplexing for multi-port CSI-RS*

The reference signals per resource block and antenna port are multiplexed into:

– frequency (FDM: Frequency Domain Multiplexing), the reference signal occupies a resource element on a subcarrier of the resource block, and the CSI-RS signals for all of the antenna ports are distributed over the resource block without interfering;

– time (TDM: Time Domain Multiplexing), the reference signal occupies one resource element on one symbol per slot, and all of the CSI-RS signals are distributed over the duration of a slot;

– code (CDM: Coding Domain Multiplexing), the same time and frequency resources are used for several antenna ports (2, 4 or 8) thanks to an Orthogonal Cover Code (OCC), which allows us to proceed to the separation of the signals reference at the receiver side.

The OCC code = Wf * Wt is defined by Table 4.8, according to the number of antenna ports (2, 4 or 8). Index i defines the code for the i-th antenna port.

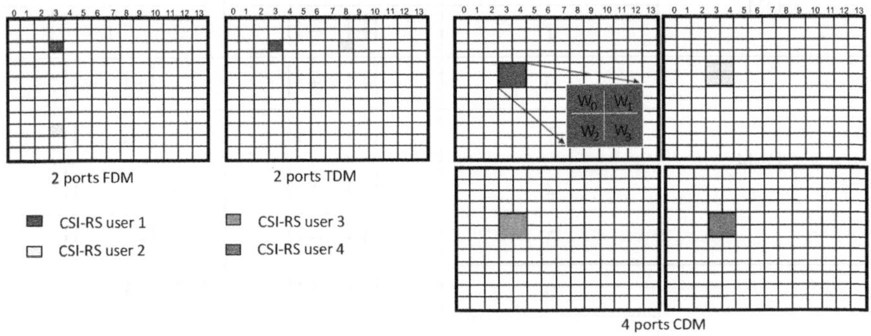

Figure 4.9. *CSI-RS multiplexing multi-ports*

4.5.2.2. *Physical DM-RS and PT-RS signal*

The DM-RS signal transmitted by the gNB base station on the downlink is associated with either the PDSCH channel, the PDCCH channel or the PBCH channel.

4.5.2.2.1. Physical DM-RS signal associated with PDSCH channel

The physical DM-RS signal is precoded with the same code as the PDSCH data. On the reception side, the DM-RS helps the receiver to estimate the channel and apply an equalization to the PDSCH signal, without needing to know the precoding matrix used for the data.

Associated with the PDSCH channel, the DM-RS is a Gold sequence generated by the following formula:

$r(n)=\frac{1}{\sqrt{2}}(1-2.c(2n))+j\frac{1}{\sqrt{2}}(1-2.c(2n+1))$, with c(n) as a Gold sequence calculated from the following two M-sequences:

$$c(n)=\left(x_1(n+N_c)+x_2(n+N_c)\right)\bmod 2$$
$$x_1(n+31)=\left(x_1(n+3)+x_1(n)\right)\bmod 2$$
$$x_2(n+31)=\left(x_2(n+3)+x_2(n+2)+x_2(n+1)+x_2(n)\right)\bmod 2$$

At the start of each subframe, the initial value c_{init} of the pseudo-random sequence is calculated from the following formula:

$$c_{init}=\left(2^{17}(14.n^{\mu}_{s,f}+l+1)(2n^{ns}_{ID}+1)+2^{17}\left\lfloor\frac{\lambda}{2}\right\rfloor+2n^{ns}_{ID}+n^{\lambda}\right)\bmod 2^{31}$$

with $n^{\mu}_{s,f}$ as the slot number in the frame and l as the OFDM symbol number in the slot. The *ns* exponent parameter is either 0 or 1. The n^{0}_{ID} and n^{1}_{ID} values are provided by the RRC layer via the *scramblingID0* and *scramblingID1* parameters or take the physical identity number of the PCI cell retrieved at the end of the time synchronization of the mobile as their value. λ is the CDM coding group and n^{λ} is 0 or 1.

The number of DM-RS symbols and the mapping of DM-RS symbols to RE are statically configured by RRC parameters, or dynamically by DCI control information. The physical DM-RS is present on each RB of the slot where the PDSCH is allocated.

In the time domain, two types of mapping are defined:

– DM-RS Mapping Type A is based on the slot time interval. The OFDM symbols carrying the DM-RS signals are fixed relative to the start of the slot and are located at position 2 or 3 of the slot;

– DM-RS Mapping Type B for which the DM-RS signal is temporally aligned with the PDSCH.

Type A allows the mobile to receive control information on the first symbol, then the DM-RS associated with the PDSCH on symbol 2 or symbol 3, the slot being entirely used to transmit data to the terminal.

Type B is used in the case where the PDSCH data is transmitted over a few symbols (mini slot over 2, 4 or 7 symbols). This configuration corresponds to low latency communications.

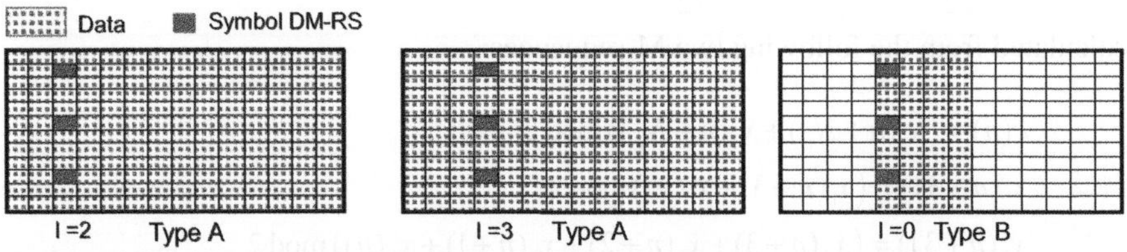

Figure 4.10. *Three examples of the DM-RS signal and mapping on antenna port 0*

The DM-RS is transmitted on one symbol (Single Symbol) or on two consecutive symbols (Double Symbol) per slot depending on the *maxlength* parameter transmitted by the RRC signaling.

At a minimum, one DM-RS symbol per UE is transmitted per slot, but depending on the RRC configuration, additional DM-RS symbols per slot may be added (*dmrs-AdditionalPosition*) in the time domain.

For high speed terminal velocity, the gNB can add three additional DM-RS signals per slot in order to help the terminal take the temporal variation of the propagation channel into account.

The reservation of resource elements for the DM-RS allows the reference signal to be transmitted on several antenna ports simultaneously without interference.

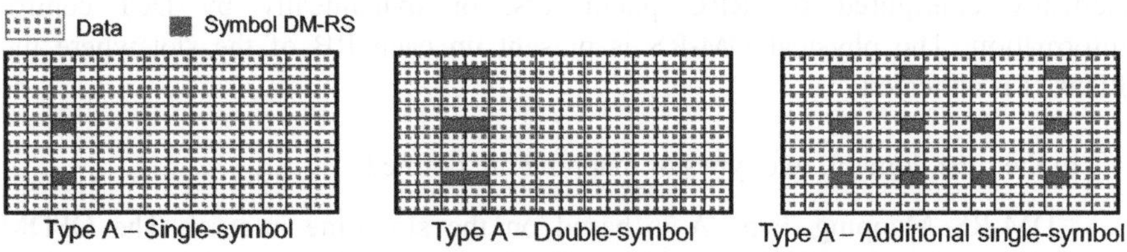

Figure 4.11. *Additional DM-RS mapping on antenna port 0*

In addition, to support different MIMO transmission modes, it is necessary to simultaneously transmit the DM-RS signals associated with the PDSCH channel, without interfering with each other through frequency division multiplexing (FDM) or through multiplexing by code (CDM).

In the frequency domain, two types of configuration are defined to indicate the frequency density:

– type 1 configuration: six subcarriers per RB and per antenna port at the location k = 4n + 2k '+ Δ, k' and Δ take the value 0 or 1;

– type 2 configuration: four subcarriers per RB and per antenna port, with two contiguous REs at the location k = 6n + k '+ Δ.

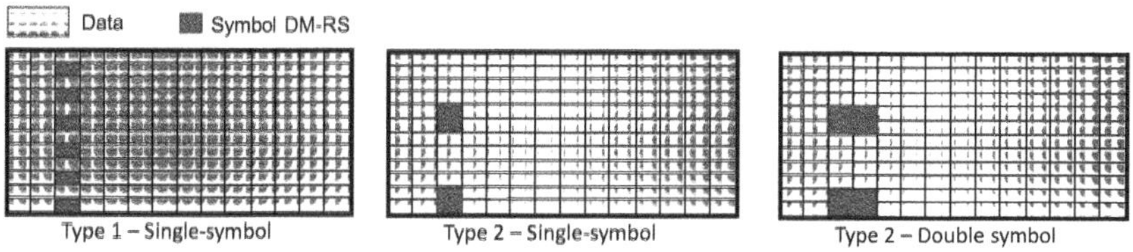

Figure 4.12. *Mapping of DM-RS in the frequency domain*

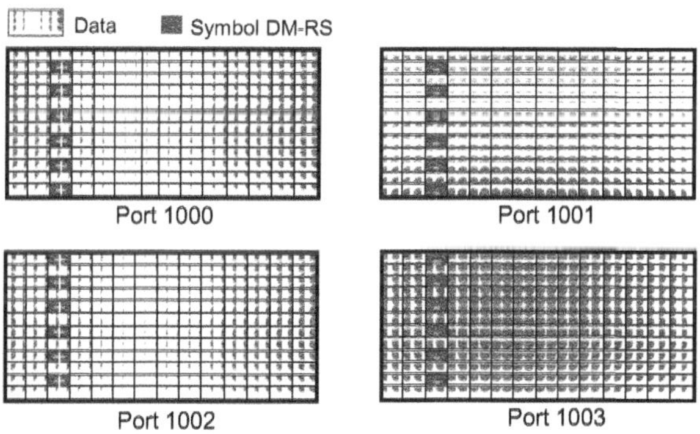

Figure 4.13. *Multi-port mode (4 DM-RS) in the case of type 1 single-symbol configuration*

By implementing frequency shift or code multiplexing:

– type 1 allows four DM-RS signals to be multiplexed (Figure 4.13) if a single symbol is mapped (single-symbol) and eight DM-RS in the double-symbol case are transmitted on antenna ports 1000–1007;

– type 2 allows 6 DM-RS signals to be multiplexed if a single symbol is mapped (single-symbol) and 12 DM-RS signals in the double-symbol case are transmitted on antenna ports 1000–1011.

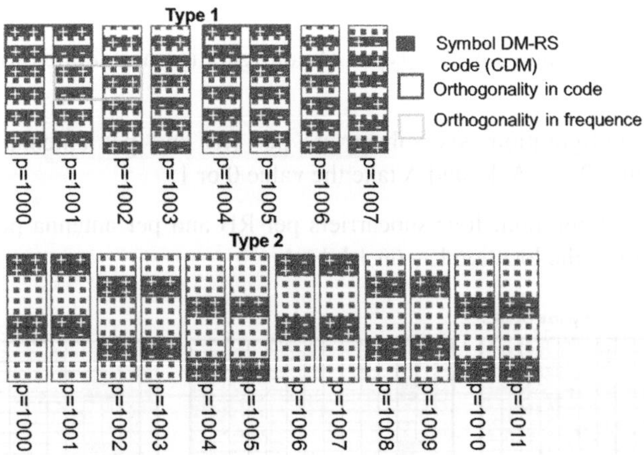

Figure 4.14. *Multi-port double-symbol (Table 7.4.1.1.2-1/2 TS 38.211)*

Orthogonality on each antenna port is obtained in the frequency domain by transmitting the reference signal on different subcarriers, or by using the OCC code on the same subcarriers (see Table 4.8).

Orthogonality in code is such that the scalar product of two codes is zero.

4.5.2.2.2. Physical PT-RS signal

The physical PT-RS signal is defined by the following formula:

$r_k = r(2M + k')$, with r being the DM-RS.

The PT-RS is optional, the terminal is informed of the presence of the PT-RS during radio link reconfiguration (RRC signaling) via the *timedensity* parameter carried by the *PTRS-DownlinkConfig* data structure.

The physical PT-RS is a copy of the DM-RS; it should not be mapped to the resource elements used for the DM-RS and CSI-RS physical signals.

In the time domain, the PT-RS is repeated every L symbols, taking the DM-RS as reference.

In the frequency domain, the PT-RS is transmitted with a density of one RB out of 2, or one RB out of 4.

4.5.2.2.3. The physical DM-RS signal associated with the PDCCH channel

Associated with the PDCCH, the DM-RS is a Gold sequence generated by the following formula:

$r(n)=\frac{1}{\sqrt{2}}(1-2.c(2n))+j\frac{1}{\sqrt{2}}(1-2.c(2n+1))$, with c(n) as the Gold sequence extracted from the following two M-sequences:

$$
\begin{aligned}
&c(n)=\left(x_1(n+N_c)+x_2(n+N_c)\right)\bmod 2\\
&x_1(n+31)=\left(x_1(n+3)+x_1(n)\right)\bmod 2\\
&x_2(n+31)=\left(x_2(n+3)+x_2(n+2)+x_2(n+1)+x_2(n)\right)\bmod 2
\end{aligned}
$$

At the start of each subframe, the initial value c_{init} of the pseudo-random sequence is calculated from the following formula:

$$c_{init}=\left(2^{17}(14.n^{\mu}_{s,f}+l+1)(2n_{ID}+1)+2n_{ID}\right)\bmod 2^{31}$$

with $n^{\mu}_{s,f}$ as the slot number in the frame and l as the position of the slot OFDM in the slot. The parameter n_{ID} corresponds to the identity of the PCI cell. The PCI is calculated by the mobile at the end of the time synchronization, or can be provided by the RRC layer via the *pdcch_DMTS_ScramblingID* parameter.

The mapping to the resource elements (k, l), with k as the subcarrier and l as the symbol number, is defined by the following formula:

k=12n+4k'+1 with k'=0,1 or 2.

The reference point corresponds to the first subcarrier on which the set of CORESET control resource elements (Control RE Set) are transmitted. This information is transmitted to the terminal in an SIB message, or during the configuration of the radio link of the BWP band.

4.5.2.2.4. The physical DM-RS signal associated with the PBCH channel

Associated with the PBCH, the DM-RS is a Gold sequence generated by the following formula:

$r(n)=\frac{1}{\sqrt{2}}(1-2.c(2n))+j\frac{1}{\sqrt{2}}(1-2.c(2n+1))$, with c(n) as the Gold sequence extracted from the following two M-sequences:

$$c(n)=\left(x_1(n+N_c)+x_2(n+N_c)\right)\bmod 2$$
$$x_1(n+31)=\left(x_1(n+3)+x_1(n)\right)\bmod 2$$
$$x_2(n+31)=\left(x_2(n+3)+x_2(n+2)+x_2(n+1)+x_2(n)\right)\bmod 2$$

At the start of each subframe, the initial value c_{init} of the pseudo-random sequence is calculated from the following formula:

$$c_{init}=\left(2^{11}\left(\bar{i}_{SSB}+1\right)\left(\left\lfloor\frac{N_{ID}^{cell}}{4}\right\rfloor+1\right)+2^{6}\left(\bar{i}_{SSB}+1\right)+\left(N_{ID}^{cell}\bmod 4\right)\right)$$

with $\bar{i}_{SSB}$, which depends on the number of the slot SSB in an SSB burst.

The mapping of the DM-RS signal associated with the PBCH channel is shown in Table 4.7.

4.5.2.3. *The physical PRS signal*

The physical PRS was introduced in Release 9.

The physical PRS is used by the mobile for the implementation of the OTDOA function.

The OTDOA function is based on the measurement, by the mobile, of the time difference between the reception of the physical PRS signal from the reference cell.

The location of the mobile is obtained from three measurements taken on three geographical neighbor cells.

The physical PRS is associated with the antenna port 5000.

The physical PRS should not be mapped to the resource elements used for the physical PSS, SSS and PBCH.

The PRS physical signal is a Gold sequence generated by the following formula:

$r(n)=\frac{1}{\sqrt{2}}(1-2.c(2n))+j\frac{1}{\sqrt{2}}(1-2.c(2n+1))$, with c(n) as the Gold sequence extracted from the following two M-sequences:

$$c(n)=\left(x_1(n+N_c)+x_2(n+N_c)\right)\bmod 2$$
$$x_1(n+31)=\left(x_1(n+3)+x_1(n)\right)\bmod 2$$
$$x_2(n+31)=\left(x_2(n+3)+x_2(n+2)+x_2(n+1)+x_2(n)\right)\bmod 2$$

At the start of each subframe, the initial value c_{init} of the pseudo-random sequence is calculated from the following formula:

$$c_{init}=\left(2^{22}\left\lfloor\frac{n_{ID,seq}^{PRS}}{1024}\right\rfloor+2^{10}(14.n_{s,f}^{\mu}+l+1)\left(2\left(n_{ID,seq}^{PRS}\bmod 1024\right)+1\right)+\left(n_{ID,seq}^{PRS}\bmod 1024\right)\right)\bmod 2^{31}$$

$n_{ID,seq}^{PRS}$ is a random number between 0 and 4095 which is transmitted by the RRC layer in the *dl-PRS-SequenceIDr-16* parameter and l corresponds to the number of OFDM symbols in the slot.

A PRS burst is made up of several PRS blocks. A PRS block consists of a PRS sequence mapped to consecutive and configurable symbols (PRS occasion).

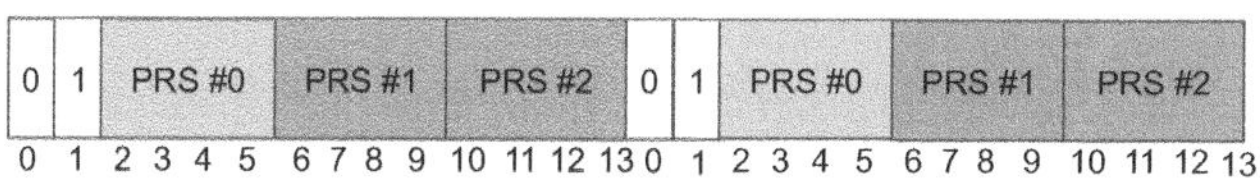

Figure 4.15. *The occasion of the PRS signal*

The PRS mapping is transmitted in an RRC message through the *dl-PRS-ResourceSet* data structure:

– in the time domain, the position of the first PRS symbol in the slot is transmitted by the *dl-PRS-ResourceSymbolOffset-r16* parameter and the number of symbols in the L_{PRS} slot is defined by the *dl-PRS-NumSymbols-r16* parameter;

– in the frequency domain, the position of the first subcarrier is transmitted by the *dl-PRS-ReOffset-r16* parameter. The PRS signal is transmitted in the form of a comb. The size of the K_{COMB} comb is passed in the *dl-PRS-CombSizeN-r16* parameter.

The PRS cannot be transmitted in the first three symbols.

The {L_{PRS}, K_{COMB}} pair defines the frequency offset of each symbol of the slot according to Table 4.9.

The pair of values {L_{PRS}, K_{COMB}} can take {2,2}, {4,2}, {6,2}, {12, 2},{4,4}, {12,4}, {6,6}, {12,6} and {12,12} as values.

K_{COMB}	Offset applied to the symbol											
	0	1	2	3	4	5	6	7	8	9	10	11
2	0	1	0	1	0	1	0	1	0	1	0	1
4	0	2	1	3	0	2	1	3	0	2	1	3
6	0	3	1	4	2	5	0	3	1	4	2	5
12	0	6	3	9	1	7	4	10	2	8	5	11

Table 4.9. *Frequency offset*

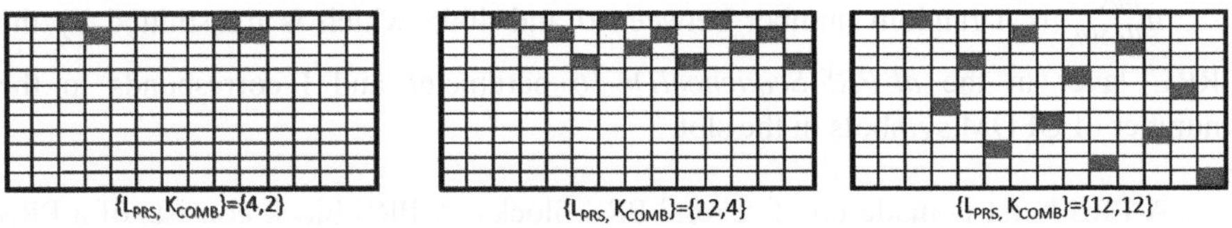

Figure 4.16. *The mapping of PRS*

4.5.2.4. *The physical RIM-RS signal*

For the TDD duplexing mode, the synchronization of gNBs is essential to prevent the downlink communications from one gNB interfering with the uplink communications between a user under the control of another gNB.

Despite the presence of a guard interval in the time domain when switching the direction of communication uplink/downlink, atmospheric conditions can contribute to the propagation of the signal over long distances.

Since the downlink signal power is higher than the uplink signal power, the downlink propagation distance is higher and the propagation delay may cause interference. This interference is called remote interference (RI).

The physical RIM-RS signal is defined from Release R.16 in order to estimate the downlink interference caused by the furthest gNB.

The physical RIM-RS is a Gold sequence generated by the following formula:

$r(n) = \frac{1}{\sqrt{2}}(1 - 2.c(2n)) + j\frac{1}{\sqrt{2}}(1 - 2.c(2n+1))$, with c(n) as the Gold sequence extracted from the following two M-sequences:

$$c(n) = (x_1(n + N_c) + x_2(n + N_c)) \bmod 2$$
$$x_1(n+31) = (x_1(n+3) + x_1(n)) \bmod 2$$
$$x_2(n+31) = (x_2(n+3) + x_2(n+2) + x_2(n+1) + x_2(n)) \bmod 2$$

At the start of each subframe, the initial value c_{init} of the pseudo-random sequence is calculated from the following formula:

$$c_{init} = (2^{10} f(n_t^{RIM}) + n_{SCID}) \bmod 2^{31}$$

The RIM-RS signal is periodically transmitted. The periodicity can be semi-static and determined by the gNB entity.

The transmission period of the RIM-RS must be a multiple of the duration of the TDD DL/UL pattern and the same for all gNBs.

The frame duration is 10 ms; the cell-specific configuration is defined by one or two profiles (pattern) transmitted to the mobile via an RRC message (see section 4.2.1).

During this periodicity, several occasions of RIM-RS transmissions are configurable to distinguish the RIM-RS resources by radio network:

– the physical RIM-RS signal is transmitted on each DL-UL switch;

– the physical RIM-RS signal is transmitted during UL-DL switching on one of the two profiles.

For each gNB radio node, several RIM-RS configurations share the same resources in the frequency domain.

4.5.3. *Physical control and data channels*

4.5.3.1. *The physical PDCCH channel*

The physical PDCCH channel transmits the downlink control information (DCI) for one or more UE mobiles. DCI messages are used to:

– define the allocation of resources, the modulation and coding scheme, for the data contained in the physical downlink channel (PDSCH) and in the physical uplink channel (PUSCH);

– control the transmission power of the PUCCH and PUSCH.

For each subframe, the PDCCH is transmitted in a grouping of time-frequency resources called CORESET.

CORESET is an aggregation of the Control Channel Element (CCE).

Each CCE occupies six Resource Element Groups (REG).

Each REG corresponds to an RB on an OFDM symbol, a CCE comprises 72 REs, 18 of which carry the DM-RS associated with the PDCCH and 54 RE for the physical PDCCH. Each symbol is encoded by QPSK modulation.

The REGs forming an CCE can be interleaved to provide frequency diversity.

In the frequency domain, the CORESET consists of a block of $N_{RB}^{CORESET}$ resources defined by the *CORESET-freq-dom* parameter. This parameter is transmitted to the mobile by RRC signaling.

In the time domain, the CORESET is spread over $N_{SYMB}^{CORESET}$ symbols defined by the *CORESET-time-dur* parameter and carried by the RRC signaling.

The number of REG forming the CORESET is defined by the $N_{REG}^{CORESET}$ parameter.

Unlike the LTE radio interface, the physical NR-PDCCH channel is not transmitted over the entire frequency band, but only in a BWP sub-partition. The physical NR-PDCCH channel is similar to the physical LTE E-PDCCH channel.

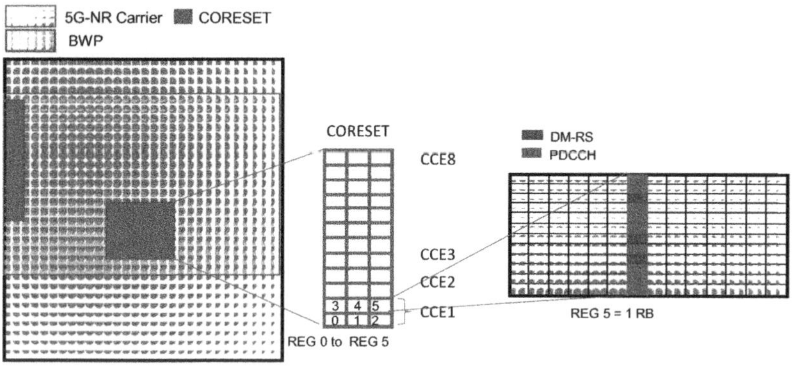

Figure 4.17. *The PDDCH channel in a slot and at PRB*

The physical PDCCH and the physical DM-RS associated with the PDCCH are transmitted in an aggregation of contiguous CCEs (1, 2, 4, 8 or 16 aggregations). The aggregation number is defined by the aggregation level parameter.

The number of bits that can be transmitted in the PDCCH is equal to 54 RE x 2 bits x Aggregation level, i.e. 108, 216, 432, 864, 1728 bits for the aggregation of 1, 2, 4, 8 or 16 CCE respectively.

The DCI information is transmitted in the CORESET. This is information relating to the physical layer and the MAC layer, called L1/L2 control signaling.

The CCE is the smallest unit in the search space for the blind decoding of DCI information.

The search space is a subset of resource elements within CORESET. This subset is scanned by the UE mobiles to detect the presence of DCI information intended for it.

There are two types of search space:

– the Common Search Space (CSS);

– the UE Search Space (USS, Specific Search Space).

The search for the presence of DCI information is carried out using an RNTI (Radio Network Temporary Identifier).

RNTIs are used to identify information which is dedicated to a UE in the cell or a group of UEs. RNTI is a 16-bit identifier and its value depends on the type of RNTI.

The C-RNTI (Cell RNTI) and CS-RNTI (Configured Scheduling RNTI) are two unique identifiers used to identify the RRC Connection and scheduling, which are dedicated to a particular UE. C-RNTI is used to allocate uplink grants, downlink assignments, etc. to a UE. C-RNTI is used by gNB to differentiate the uplink transmissions (e.g. PUSCH, PUCCH) of a UE from others.

The SI-RNTI (System Information RNTI) identifier is used to broadcast system information (SIB). SI-RNTI is like a broadcast RNTI and the value is 0xFFFF.

The P-RNTI (Paging RNTI) identifier is used to broadcast a notification relating to paging information. P-RNTI is like a broadcast RNTI and the value is 0xFFFE.

The RA-RNTI (Random Access RNTI) identifier is used to retrieve the information transmitted during the random access procedure.

The TC-RNTI (Temporary Cell RNTI) is also used during the random procedure to resolve contention.

The TCP-RNTI (Transmit Power Control RNTI) is used for uplink control power. Three types, TCP-PUSCH-RNTI, TCP-PUCCH-RNTI and TCP-SRS-RNTI, make it possible to retrieve the information corresponding to the power control of the PUSCH traffic channels, PUCCH control and the SRS reference signal, respectively.

The SP-CSI-RNTI (Semi Persistent CSI Reporting) identifier enables or disables the resending of CSI measurements on the PUSCH upstream data channel.

The SFI-RNTI (Slot Format Indicator) identifier is used to notify the transmission format (DL, UL or flexible) of a slot for one or more UEs.

The MCS-C-RNTI (Modulation Coding Scheme) identifier is used to dynamically configure the coding and modulation scheme for unicast IP data.

The INT-RNRI (RNTI interrupt) identifier is used to notify the mobile of a preemption of its resources.

DCI formats 0-0 and 0-1 allocate the resources for the physical PUSCH channel. The size of the 0-0 format (fallback format) is standard and cannot be configured to allow the allocation of common resources, regardless of the category of mobile. The 0-1 format is used to transmit the entire configuration for specific terminals.

The DCI format 1-0 introduces the allocation of resources in the physical PDSCH channel, or information relating to random access, paging and MIB system information. The size of the DCI format 1-0 (fallback format) is limited and concerns all 5G terminals by default. The informational data is limited to the common characteristics for the allocation of PDSCH resources.

The DCI format 1-1 introduces the allocation of all resources concerning the physical PDSCH channel. The size of the DCI message depends on the complete configurations (aggregation of carriers, etc.).

The DCI format 2-0 introduces dynamic allocation for a set of terminals by configuring the direction of the SFI slots.

The DCI format 2-1 introduces a dynamic preemption of a resource initially allocated to the mobile and preempted following an interruption for a priority mobile.

The DCI formats 2-2 and 2-3 are used for power control of the PUSCH/PUCCH and SRS channels respectively.

DCI Format	Usage	RNTI
Format 0_0	PUSCH scheduling	C-RNTI, TC-RNTI, CS-RNTI, MCS-C-RNTI
Format 0_1	PUSCH scheduling	C-RNTI, CS-RNTI, MCS-C-RNTI, SP-CSI-RNTI
Format 1_1	PDSCH scheduling	C-RNTI, TC-RNTI, CS-RNTI, MCS-C-RNTI, P-RNTI, SI-RNTI, RA-RNTI
Format 1-1	PDSCH scheduling	C-RNTI, CS-RNTI, MCS-C-RNTI
Format 2-0	Notification of slot format	SFI-RNTI
Format 2-1	Notification of an interruption of data transmission	INT-RNTI
Format 2-2	Power control for PUSCH or PUCCH	TPC-PUCCH-RNTI, TPC-PUSCH-RNTI
Format 2-3	Power control for SRS	TPC-SRS-RNTI

Table 4.10. *Radio identifier associated with the DCI format*

The search space is indicated by a set of contiguous CCEs. The UE monitored search space to decode DCI scrambled with RNTI and read PDCCH. DCI carried scheduling assignments/grants relating to a certain component carrier. A blind search is performed when the UE mobile does not have all of the information.

The CSS (Common Search Space) defines the set of DCI formats that each UE mobile must scan, for the presence of common control information.

The UE USS defines the set of DCI formats that the UE must scan for dedicated control information. The USS is indicated to the mobile UE by an RRC signaling message.

4.5.3.2. *Physical PDSCH channel*

The physical PDSCH channel carries data sharing the capacity on a time and frequency basis.

The PDSCH transmits the DL-SCH and PCH transport channels.

The DL-SCH contains the data specific to the UE terminals corresponding to the traffic data (unicast IP packets) and to the RRC (Radio Resource Control) control data, and contains SIB system information.

The DL-SCH is in the form of one or two transport blocks.

The PCH contains the RRC control data corresponding to the paging message.

The PDSCH uses a modulation and coding scheme depending on the transmission channel with each UE terminal.

The 5G-NR interface only offers one transmission mode for the PDSCH. However, this mode allows up to 8 transmission layers to be exploited on the 12 antenna ports from 1000 to 1011.

The bit rate available to a user is a function of the modulation scheme, the performance of the encoder, the allocated frequency band and the implementation of spatial diversity (MIMO).

The number of spatial layers exploited by the MIMO technology, the valence of the modulation (QPSK, 16 QAM, 64 QAM, 256 QAM), J the number of aggregated

5G-NR bands, the coding efficiency and f (j) the factor corresponding to the TDD configuration on the jth 5G-NR band (j = 1 if all the traffic is in DL, 0.8 or 0.75 or 0.4 depending on the configurations chosen between the upstream and downstream direction)):

$$DataRate = 10^{-6}\sum_{j=1}^{J}\left(v_{layers}^{(j)}.Q_m^{(j)}.f^{(j)}.R_{\max}.\frac{N_{PRB}^{BW(j),\mu}.12}{T_s^{\mu}}.(1-OH^{(j)})\right)$$

with $T_s^{\mu} = \frac{10^{-3}}{14.2^{\mu}}$, as the symbol duration for the numerology μ and $N_{PRB}^{BW(j),\mu}$, as the number of PRB in the jth 5G-NR band.

The OH parameter corresponds to the overhead of the signaling message. The current values are OH = 0.14 for the FR1 band and 0H = 0.18 for the FR2 band.

There are two types of resource allocation in the time domain for the downlink named Type A and Type B (see section 4.5.2.2.1). Type A is an allocation on an entire slot, type B is an allocation for a mini slot.

The PDSCH resource allocation in the time domain is indicated by the value of the time domain resource assignment field carried by the DCI message. The allocation table (Table 4.10) defines the offset value of the slot K0 carrying the PDSCH message with respect to the DCI message received, the length L of the PDSCH message and the position S of the first symbol carrying the PDSCH message (SLIV indicator: start and length indicator)

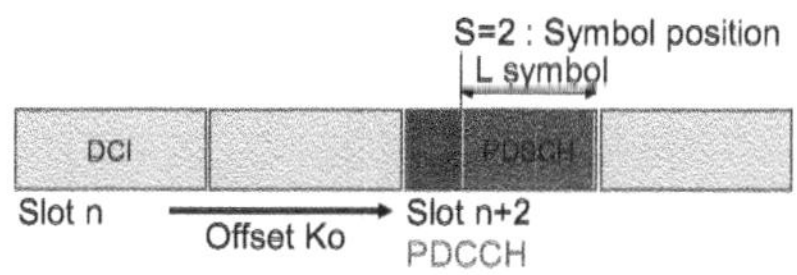

Figure 4.18. *The allocation of the PDSCH message in the time domain (SLIV)*

The specification offers three types of resource allocation in the time domain. Configuration A is presented in Table 4.11, and configurations B and C can be retrieved from the specification 3GPP TS38.214 Table 5.1.2.1.1-4/5.

In the frequency domain, there are two types of resource allocation for the downlink.

Row index	*dmrs-TypeA-Position*	PDSCH mapping type	K_0	S	L
1	2	Type A	0	2	12
	3	Type A	0	3	11
2	2	Type A	0	2	10
	3	Type A	0	3	9
3	2	Type A	0	2	9
	3	Type A	0	3	8
4	2	Type A	0	2	7
	3	Type A	0	3	6
5	2	Type A	0	2	5
	3	Type A	0	3	4
6	2	Type B	0	9	4
	3	Type B	0	10	4
7	2	Type B	0	4	4
	3	Type B	0	6	4
8	2,3	Type B	0	5	7
9	2,3	Type B	0	5	2
10	2,3	Type B	0	9	2
11	2,3	Type B	0	12	2
12	2,3	Type A	0	1	13
13	2,3	Type A	0	1	6
14	2,3	Type A	0	2	4
15	2,3	Type B	0	4	7
16	2,3	Type B	0	8	4

Table 4.11. *The allocation of PDSCH in the time domain (TS 38.214, Table 5.1.2.1.1-2) for configuration A*

Type 0 resource allocation consists of allocating groups of consecutive PRBs in frequency, called RBG (Resource Block Group). Each RBG allocated to the mobile UE is signaled by a message, each bit of which corresponds to the allocation or not of the RBG (bitmap allocation).

The number of RBs in an RBG depends on the size of the BWP bandwidth (Table 4.12).

Number of PRB in BWP	Configuration 1	Configuration 2
1–36	2	4
37–72	4	8
73–144	8	16
145–275	16	16

Table 4.12. *The length of RBG in function of the bandwidth (TS 38.214, Table 5.1.2.2.1-1)*

Configuration 1 or 2 is transmitted by the *RBGSize* field contained in the PDSCH-Config structure of the RRC message.

Type 1 resource allocation allocates a set of Virtual Resource Blocks (VRB).

Type 1 allocation involves allocating a starting block of resources and a number of VRBs. This information is encoded in a Resource Indication Value (RIV), transmitted in a DCI message.

The VRB mapping to PRB resources can be localized (one VRB corresponds to one PRB) or distributed (frequency hopping).

4.6. Transmission in uplink

The communication on the uplink channel is based on either the OFDM signal or on the DFT-S-OFDM signal (see section 4.1.1).

The DFT-S-OFDM signal reduces energy consumption and is obtained by applying an additional Fourier transform in the transmission chain, compared to the OFDM transmission chain.

The reference signals associated with this signal will be CAZAC sequences (Constant Amplitude Zero AutoCorrelation). CAZAC sequences allow energy consumption to be reduced (constant amplitude).

Zadoff–Chu sequences are CAZAC sequences, complex sequences whose versions obtained by cyclic shifts are orthogonal.

The Zadoff–Chu sequences of length M are defined by a root u forming a base sequence, from which orthogonal sequences are derived by cyclic shifting.

The formulation of a Zadoff–Chu sequence is as follows: $z_i^u = e^{-j\frac{\pi ui(i+1)}{M}}$

In the specifications of 3GPP Release R.16, CAZAC sequences are named $r_{u,v}$:

– 30 groups of CAZAC sequences (u = {0,… 29}) of length M are defined;

– in each group, the basic sequence corresponds to v = 0.

If the length of the sequence is greater than or equal to 36 (equivalent to 3 RB), then $r_{u,0}(n) = z_n^u$.

If the length of the sequence is equal to {6,12,18,24}, then $r_{u,v}(n) = e^{j\varphi(n)\pi/4}$ with $\varphi(n) \in \{-3,-1,1,3\}$. The value of $\varphi(n)$ is defined by Tables 5.2.2.2-1/4 in specification TS 38.211.

If the length of the sequence is equal to 30, then $r_{u,v}(n) = e^{-j\frac{\pi(u+1)(n+1)(n+2)}{31}}$.

5G-NR supports DFT-S-OFDM and OFDM in the uplink direction. When the OFDM signal is used, the reference signal is a Gold sequence.

The Gold sequence r(n) is generated from two M-sequences (pseudo-random sequences)

$$r(n) = \frac{1}{\sqrt{2}}(1 - 2.c(2n)) + j\frac{1}{\sqrt{2}}(1 - 2.c(2n+1))$$

with c(n) extracted from 2 M-sequences:

$$c(n) = (x_1(n + N_c) + x_2(n + N_c)) \bmod 2$$
$$x_1(n+31) = (x_1(n+3) + x_1(n)) \bmod 2$$
$$x_2(n+31) = (x_2(n+3) + x_2(n+2) + x_2(n+1) + x_2(n)) \bmod 2$$

4.6.1. *Physical reference signals*

4.6.1.1. *Physical SRS signal*

The physical SRS (Sounding RS) signal is a reference signal transmitted by the UE. The gNB performs a coherent demodulation of the received signal and assesses the transfer function of the radio channel. The received SRS is also used to measure the quality of the radio link (CSI).

The CSI measurement is used by the base station for the sequencing and adaptation of the specific link to the UE.

In TDD, by exploiting the channel reciprocity, the transmitter can estimate the downlink channel from the sounding on the uplink channel. The SRS physical signal, configured at the mobile level, is limited to four antenna ports.

The SRS signal is a CAZAC Zadoff–Chu sequence.

A cyclic shift in the time domain (phase rotation of the Zadoff–Chu sequence) allows the physical SRS to be multiplexed between each antenna port.

The position in the time domain of the SRS is located in the last six symbols of the slot (symbol 8 to symbol 13) and is spread over one, two or four consecutive OFDM symbols (see Figure 4.19).

The SRS position is defined by $l = 13 - l_{0ffset}$ with $l_{0ffset} = \{0,1,2,3,4,5\}$.

The frequency location of the SRS forms a comb: the mobile uses a subcarrier on K_{TC}, K_{TC}, taking the value 2, 4 or 8 (see Figure 4.19).

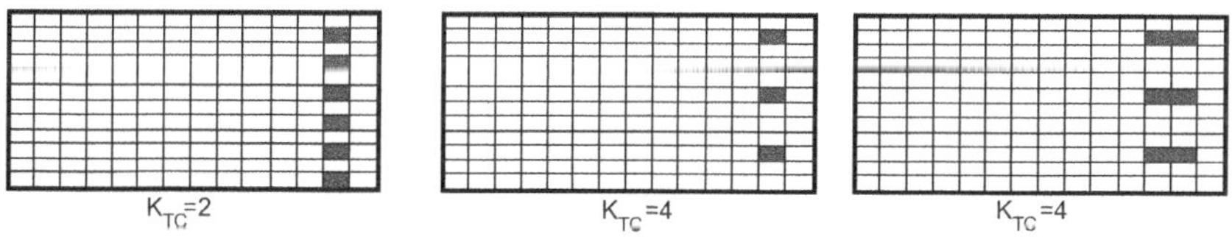

Figure 4.19. *SRS signal mapping (one or two symbols)*

The SRS signals transmitted by different mobiles are multiplexed in the frequency domain. If $K_{TC} = 2$, two SRS can be multiplexed; if $K_{TC} = 4$, four SRS can be multiplexed.

4.6.1.2. *The DM-RS signal associated with the PUSCH channel*

The physical DM-RS signal is transmitted with the PUSCH signal. The same precoding is applied to the PUSCH channel and DM-RS channel to help the receiver to demodulate the signal, without needing to know the precoding matrix used for the data.

Associated with the PUSCH, the DM-RS is either a Gold sequence or a CAZAC sequence, depending on the shape of the signal (OFDM or DFT-S-OFDM, respectively).

When the DM-RS associated with the PUSCH is based on DFT-S-OFDM modulation, the DM-RS reference sequence is a CAZAC sequence generated by the formula: $r(n,) = r_{u,v}^{(\alpha_i,\delta)}(n)$.

At the start of each subframe, the initial value c_{init} of the pseudo-random sequence is calculated from the following formula:

$$c_{init} = \left(2^{17}(14.n_{s,f}^{\mu} + l + 1)(2n_{ID}^{ns} + 1) + 2_{ID}^{ns} + n^{\lambda}\right) \bmod 2^{31}$$

When the DM-RS associated with the PUSCH is based on OFDM modulation, the DM-RS reference sequence is a Gold sequence generated by the formula:

$$r(n) = \frac{1}{\sqrt{2}}(1 - 2.c(2n)) + j\frac{1}{\sqrt{2}}(1 - 2.c(2n+1))$$

At the start of each subframe, the initial value c_{init} of the pseudo-random sequence is calculated from the following formula:

$$c_{init} = \left(2^{17}(14.n_{s,f}^{\mu} + l + 1)(2n_{ID}^{ns} + 1) + 2^{17}\left\lfloor\frac{\lambda}{2}\right\rfloor 2_{ID}^{ns} + n^{\lambda}\right) \bmod 2^{31}$$

with $n_{s,f}^{\mu}$ as the number of slots in the frame and 1 as the number of OFDM symbols in the slot.

The parameter ns is equal to 0 or 1. The values n_{ID}^{0} and n_{ID}^{1} are either given by the RRC layer with the *scramblingID0* and *scramblingID1* parameters, or take the value of the physical identity of the PCI cell. The PCI identity is recovered at the end of the time synchronization of the mobile. λ is the CDM coding group, taking the value of 0.1 or 2 and n^{λ} is equal to 0 or 1.

The number of DM-RS symbols and mapping to resource elements (RE) are statically configured by RRC parameters, or dynamically by control information (DCI). The physical DM-RS is present on each RB of the slot where the PUSCH is allocated.

The mapping type is used to indicate the position S of the first OFDM symbol of the DM-RS and the number L of symbols:

– DM-RS Mapping Type A is based on the slot time interval. L is the time between the first OFDM symbol of the slot and the last OFDM symbol of the PUSCH;

– DM-RS Mapping Type B, L is the duration of the allocated PUSCH resources. If frequency hopping is off, L is the duration of the hop. The DM-RS is time aligned with the PUSCH.

The position of the DM-RS associated with the PUSCH is determined by the Pos value.

The mapping is transmitted in the DCI format 0-0 or 0-1 message.

Like the mapping for the uplink, Type B mapping is used for mini-slots.

At a minimum, one DM-RS symbol per UE is transmitted per slot, but depending on the RRC configuration, additional DM-RS symbols per slot may be added (*dmrs-AdditionalPosition*) in the time domain.

The DM-RS is transmitted on one symbol (Single Symbol) or on two symbols (Double Symbol) per slot, depending on the *maxlength* parameter transmitted via the RRC signaling in the *DMRS-UplinkConfig* field of the PUSCH-Config structure.

In the case of multi-antenna ports (two or four antenna ports), the DM-RS on each port is coded (CDM: Code Division Multiplex).

In the frequency domain, two types of configuration are defined:

– type 1 configuration: six subcarriers per RB and per antenna port at the location $k = 4n + 2k' + \Delta$, k' and Δ take the value 0 or 1;

– type 2 configuration: four subcarriers per RB and per antenna port, with two contiguous REs at the location $k = 6n + k' + \Delta$.

When the DM-RS associated with the PUSCH channel relies on the DFT-S-OFDM modulation, type 1 and type 2 configurations may apply.

When the DM-RS associated with the PUSCH relies on the OFDM modulation, only the type 1 configuration applies.

4.6.1.3. *The DM-RS signal associated with the PUCCH channel*

The physical DM-RS signal associated with the PUCCH channel is used for the estimation and synchronization of the physical PUCCH channel.

The physical PUCCH supports five formats, numbered from 0 to 4.

The characteristics of the DM-RS associated with the PUCCH are correlated to the format of the PUCCH.

The DM-RS is associated with the PUCCH for formats 1, 2, 3 and 4.

The DM-RS r_{PUCCH} signal generated corresponds to the following formula:

$$r_{PUCCH} = w_l(m).r$$

where $w_l(m)$ is an orthogonal sequence applicable only for formats 1 and 2 of the physical PUCCH channel; it allows code multiplexing of several mobiles.

(m) references the OFDM symbol number assigned to the physical DM-RS signal.

The sequence is verified:

– for formats 1, 3 and 4 by the CAZAC sequence $r = r_{u,v}^{(\alpha,\delta)}$;

– for format 2, by the following Gold sequence:

$$r(n) = \frac{1}{\sqrt{2}}(1-2.c(2n)) + j\frac{1}{\sqrt{2}}(1-2.c(2n+1)) \text{ with}$$

$$c(n) = (x_1(n+N_c) + x_2(n+N_c)) \bmod 2$$
$$x_1(n+31) = (x_1(n+3) + x_1(n)) \bmod 2$$
$$x_2(n+31) = (x_2(n+3) + x_2(n+2) + x_2(n+1) + x_2(n)) \bmod 2$$

At the start of each subframe, the initial value c_{init} of the pseudo-random sequence is calculated from the following formula:

$$c_{init} = \left(2^{17}(14.n_{s,f}^{\mu} + l + 1)(2n_{ID}^{0} + 1) + 2n_{ID}^{0}\right) \bmod 2^{23}$$

The resource elements (k, l) of the physical DM-RS are located on the time–frequency resources allocated to the PUCCH.

The physical PUCCH channel is defined by a short (Short PUCCH) or long (Long PUCCH) structure.

The Short PUCCH physical channel is transmitted in the time domain on the last symbol or the last two symbols. The physical DM-RS signal and UCI information are frequency multiplexed (FDM) with a density of ½, ¼ or 1/6.

The Long PUCCH physical channel is transmitted in the frequency domain on one or more PRBs and in the time domain on a set of symbols in a slot. The physical DM-RS signal and UCI information are frequency multiplexed (TDM).

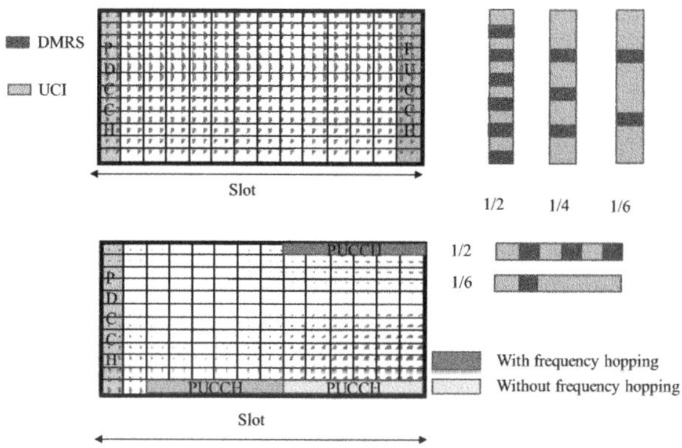

Figure 4.20. *The DM-RS signal mapping associated with the short and long PUCCH channels*

4.6.1.4. *The physical PT-RS signal*

The PT-RS signal is optional. The mobile is informed of the presence of the PT-RS signal during radio link reconfiguration (RRC signaling), via the *timedensity* parameter carried by the *PTRS-UplinkConfig*.

The physical PT-RS is a copy of the DM-RS associated with the PUSCH channel; it must not be mapped to the resource elements used for the physical DM-RS signals.

In the time domain, the PT-RS is repeated every L symbols, taking the DM-RS as reference.

In the frequency domain, the PT-RS is transmitted on the same subcarriers as the physical DM-RS (type 1 or type 2 configuration).

4.6.2. *The physical channel*

4.6.2.1. *The random physical PRACH channel*

The physical PRACH (*Physical Random Access Channel*) channel contains a preamble sent by the mobile when it has to perform a random access.

In the time domain, the signal must be transmitted during a T_{PRACH} window. T_{PRACH} window is a succession of time slots, during which the mobile has N_t^{RA} opportunities for the transmission of the physical PRACH. The duration of the PRACH sent by the mobile is defined by the value N_{dur}^{RA} and corresponds to the number of repetitions of the sequence.

In the time domain, the value of the transmission symbol of the physical PRACH channel transmitted by the mobile is calculated as follows:

$l = l_0 + n_t^{RA} N_{dur}^{RA} + 14 n_{slot}^{RA}$, l_0 is the position of the first symbol of the channel and N_{slot}^{RA} is equal to 0 or 1, depending on the spacing between the subcarriers.

In the frequency domain, the physical channel is transmitted over a set of resource blocks (RB). The start of the band and the number of PRACH opportunities allocated in the frequency domain are transmitted via the SIB2 information message by the *msg1-FrequencyStart* and *msg1-FDM* values, respectively.

Random access is the first step used by the mobile to set up a connection to the gNB entity. The mobile UE sends its connection request via the physical PRACH. If the radio node does not respond, the mobile makes a new request by increasing the transmission power according to a ramping factor (*powerRampingStep*). The number of PRACH requests is limited by the *preambleTransMax* value transmitted via the SIB2 information message.

During random access, the mobile is synchronized in transmission with the received synchronization signal. The random access procedure will estimate the radio propagation time, which is formulated by the TA (Timing Advanced) value.

The preamble consists of a Cyclic Prefix (CP) of duration T_{CP} and a sequence of duration T_{SEQ}.

The time guard T_G is equal to T_{PRACH}-(T_{CP}+ T_{SEQ}).

The guard time corresponds to the maximum round trip time (RTT) and determines the maximum value of the radius R of the cell

$$R(km)= T_G\ (\mu s)/6,6.$$

RACH sequences are Zadoff–Chu sequences over 139 bits (short sequences) or 839 bits (long sequences).

The r_{PRACH} sequences are generated by the following equation (with M=139 for a short sequence or M=839 for a long sequence):

$$z_i^u = e^{-j\frac{\pi u i(i+1)}{M}} \quad \text{and} \quad r_{PRACH}(n) = z^u_{(n+C_v) \bmod M}$$

C_v is an offset of the base sequence.

The PRACH is time and frequency multiplexed with the PUSCH channel of the other UEs.

In the frequency domain, the long sequences are transmitted in the FR1 band. PRACH sequences span 6 RB for a 1.25 kHz subcarrier spacing, or 24 RB for a 5 kHz subcarrier spacing.

In the time domain, the long sequence consists of 839 symbols. The reference duration of a slot is 1 ms.

Table 4.13 describes the format of the various preambles of the long sequences.

Format	**Subcarrier spacing**	**Number of repetition**	**Prefix duration T_{CP}**	**Preamble duration**	**Cell radius**
0	1.25 kHz	1	100 µs	800 µs	14 km
1	1.25 kHz	2	680 µs	1600 µs	75 km
2	1.25 kHz	4	15 µs	3200 µs	28 km
3	5 kHz	1	100 µs	800 µs	108 km

Table 4.13. *The structure of a long PRACH sequence*

The performance metrics for preamble detection relate to the probability of detection and the probability of false alarms.

The basic preamble can be repeated in the case of cells for which signal losses and/or interference are important.

Table 4.14 describes the format of the various short sequence preambles used in the FR1 and FR2 bands.

Format	Number of slot repetition (N_{dur}^{RA})	Prefix cyclic duration T_{CP}	Preamble duration	Number of occasion in a slot
A1	2	9.4 µs	133 µs	6
A2	4	18.7 µs	267 µs	3
A3	6	28.1 µs	400 µs	2
B1	2	7 µs	133 µs	7
B2	4	11.7 µs	267 µs	3
B3	6	16.4 µs	400 µs	2
B4	12	30.5 µs	800 µs	1
C0	1	40.4 µs	66,7 µs	7
C1	4	66.7 µs	267 µs	2

Table 4.14. *The structure of short PRACH sequences*

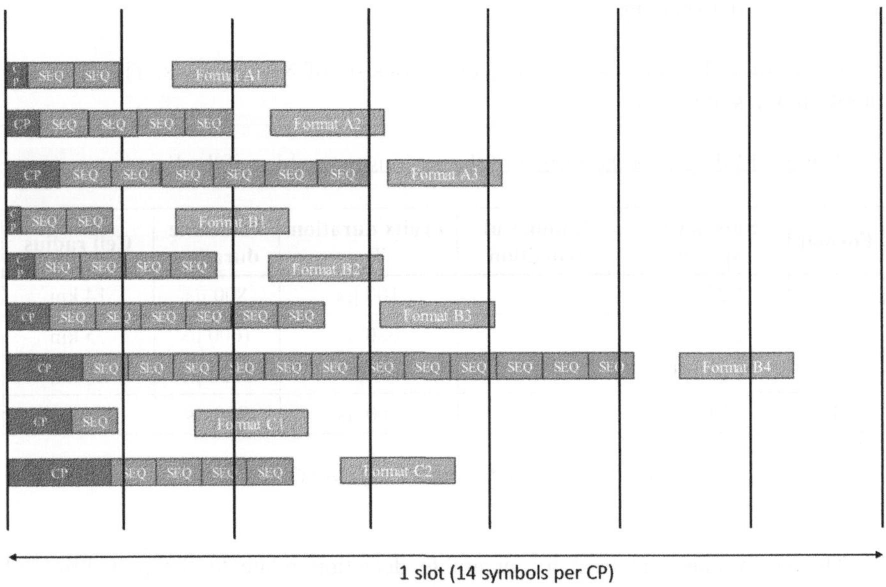

Figure 4.21. *The mapping of the DM-RS reference signal associated with short and long PUCCH*

4.6.2.2. *The physical PUCCH channel*

The physical PUCCH channel uses five types of format to transport UCI information. The type of format is chosen depending on the amount of information to be transported and the coverage of the cell.

The NR PUCCH of short duration (Short PUCCH) is transmitted on one or two symbols and can carry one or two UCI bits (format 0) or several UCI bits (format 2), by transmitting the UCI symbols on several PRBs. UCI control data is multiplexed with descending or ascending data. The modulation used is of the QPSK type.

The NR PUCCH of long duration (Long PUCCH) is transmitted over a longer duration, between 4 and 14 symbols. The modulation used is of the BPSK or QPSK type.

Format PUCCH	Number of symbol/duration	Number of UCI bits	Number of PRB
0	SHORT/1-2	1–2	1
1	LONG/4-14	1–2	1
2	SHORT/1-2	>2	1–16
3	LONG/4-14	>2	1–6, 8–10, 12, 15, 16
4	LONG/4-14	>2	1

Table 4.15. *PUCCH format*

Format 0 carries one or two UCI bits as an acknowledgment or a sequencing request. UCI 0-0 format is the default format which supports basic 5G-NR functionality and UCI 0-1 is the non-fallback format to support flexible configurations when RRC reconfiguration requests.

Format 2 takes the CSI information into account:

– the short PUCCH physical channel (short PUCCH) is transmitted on the last or the last two symbols of the slot and carries the positive ACK or negative NACK acknowledgment of the data received on the physical PDSCH channel, corresponding to the HARQ mechanism for low communication URLLC (Ultra-Reliable Low-Latency Communication);

– the long physical PUCCH channel (Long PUCCH).

The formats 1, 1a and 1b transport the UCI information relating to the scheduling request to obtain the resource on the physical PUSCH (Physical Uplink Shared Channel).

The 2, 2a and 2b formats carry UCI information relating to the status report of the signal received on the physical PDSCH and to the positive ACK or negative NACK acknowledgment.

Format 3 carries the same information as Format 1 by adapting it to the aggregation of DC radio channels.

At the uplink level, the physical PUSCH channel uses OFDM or DFT-S-OFDM modulation. In order to reduce the options on the physical PUCCH channel, formats 0, 1, 3 and 4 use the DFT-S-OFDM modulation and format 2 uses the OFDM format.

In the time domain:

– the PUCCH format 0 and format 2 physical channels are generally transmitted on the last symbol of the slot, even if another position is possible;

– the PUCCH format 1, format 3 and format 4 physical channels are transmitted on several consecutive symbols, on the same subcarrier or on two different subcarriers in the event of frequency hopping. Compared to format 3, format 4 accepts code multiplexing (CDM) between the different UEs.

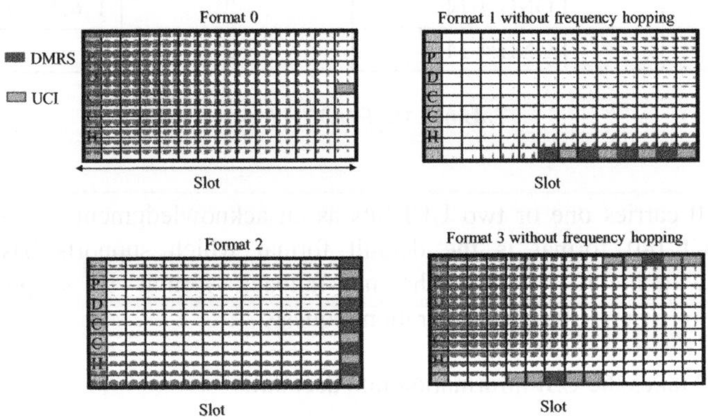

Figure 4.22. *The NR PUCCH format associated with DM-RS*

4.6.2.3. *The physical PUSCH channel*

PUSCH channel transmits the UL-SCH transport channel and can multiplex UCI control information to avoid the simultaneous transmission of the PUCCH channel.

The UL-SCH transport channel contains the traffic data (unicast IP packets) and RRC control corresponding to the following messages:

– common or dedicated control messages;

– NAS signaling transport messages (Non-Access Stratum).

UCI control information includes HI information (HARQ Indicator) and CSI (Channel State Information) reports.

There are two types of time domain resource allocations for the uplink, called Type A and Type B (see section 4.6.1.2). Type A is an allocation for an entire slot, and Type B corresponds to an allocation for a mini-slot.

For Type A mapping, the physical PUSCH channel starts at the first symbol l = 0 and the associated physical DM-RS signal is fixed at the 3rd or 4th symbol.

For Type B mapping, the physical PUSCH channel starts on any symbol. The physical DM-RS signal starts on the same symbol.

In the frequency domain, the PUSCH signal can be modulated by an OFDM signal or by a DFT-S-OFDM signal from a precoding transform.

4.7. References

All standards can be downloaded from the ETSI website: https://www.etsi.org/standards.

3GPP TS 38.211
NR - Physical Channels and Modulation;
Version 16.1.0 Release 16

3GPP TS 38.213
NR - Physical layer procedures for control;
Version 16.1.0 Release 16

3GPP TS 38.214
NR - Physical layer procedures for data;
Version 16.2.0 Release 16

3GPP TS 38.215
NR - Physical layer measurements;
Version 15.2.0 Release 15

3GPP TR 38.866
Study on remote interference management for NR;
Version 16.1.0 Release 16

3GPP 38.306
5G-NR - User Equipment (UE) radio access capabilities;
Version 15.3.0 Release 15

3GPP 38.331
NR - Radio Resource Control (RRC); Protocol specification;
Version 16.0.0 Release 16

3GPP TR 38.804
Radio Interface Protocol Aspects; Study on New Radio Access Technology;
Version 1.0.0 Release 14

3GPP R1-1900310
NR RAT-dependent DL Positioning
WG1 Ad-Hoc Meeting 1901, 21st – 25th January 2019

3GPP R1-1812616
NR PRS Design for Downlink based UE positioning

3GPP R1-1901547
Discussion on RIM-RS configuration
WG1 Meeting 96, 25th February – 1st March, 2019

5

5G-NR Radio Interface – Operations on the Frequency Bands

5.1. Operations on the frequency bands

Operations on the 5G-NR and 4G-LTE (Long-Term Evolution) frequency bands make it possible to increase cell coverage and/or improve data rate transmission on the UL (Uplink) and the DL (Downlink).

DSS (Dynamic Spectrum Sharing) consists of dividing the 4G FDD frequency band into two sub-bands, a 4G sub-band and a 5G sub-band. The 4G and 5G bandwidths are dynamically controlled by the radio node. Dynamic Spectrum Sharing is a technology that allows the deployment of both 4G LTE and 5G-NR in the same frequency band.

Carrier Aggregation (CA) consists of adding different 5G-NR radio channels. The 5G-NR specification defines two different approaches for informing the UE (User Equipment) about the scheduling for each band: a separate PDCCH for each carrier or a common PDCCH for multiple carriers (cross-carrier scheduling).

SUL (Supplementary Uplink) mode allows the addition of a low frequency radio channel. This channel is dedicated only for transmissions on the uplink direction. SUL aims to improve the uplink cell coverage. SUL mode differs from carrier aggregation since the mobile is controlled to transmit data on the PUSCH channel or signaling in the PUCCH channel either on the SUL band or on the traditional paired band, but not in both bands simultaneously.

Dual connectivity consists of operating two radio nodes simultaneously. Dual connectivity can take place on two different MR-DC (Multi-RAT Dual

Connectivity) radio interfaces such as the 5G-NSA option or on two similar NR-DC (New Radio Dual Connectivity) radio interfaces.

Figure 5.1 shows the deployment of a 5G radio node in the FR1 (Frequency Range 1) band at the frequency of 3.5 GHz on a 4G site that operates LTE in a 700 MHz band.

Operations on the different frequency bands improve coverage and throughput:

– spectrum sharing and the SUL technique allow the deployment of 5G on the FDD band at 700 MHz. The coverage is important but has weak capacity (e.g. the band is limited to 10 MHz);

– with dual connectivity, UE are controlled in the 4G band and the data rate is increased thanks to the 5G interface. With 4G connectivity at 700 MHz, the coverage is important and 5G connectivity at 3.5 GHz allows higher data rates;

– carrier aggregation increases the data rate thanks to it having more bandwidth to allocate.

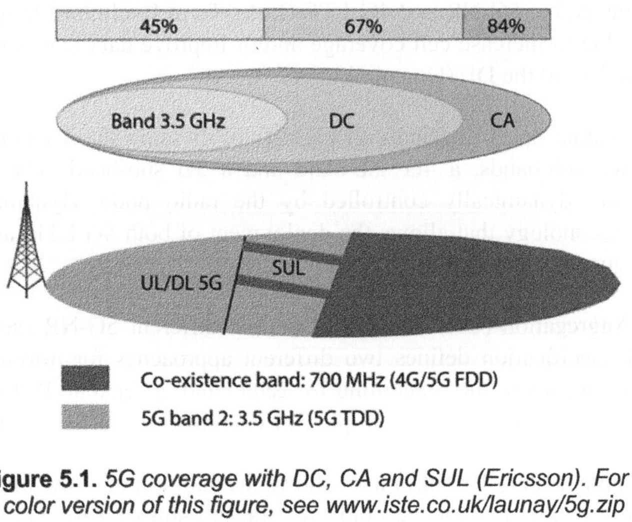

Figure 5.1. *5G coverage with DC, CA and SUL (Ericsson). For a color version of this figure, see www.iste.co.uk/launay/5g.zip*

Among all the users under 4G coverage in the cell:

– 45% of users only access 5G in the 3.5 GHz band (5G-SA). SUL mode allows mobiles to transmit on the uplink in the 700 MHz band.

– 67% of users have access to the 5G-NSA network: the control plan is carried out by the eNB base station on the LTE interface and the dual connectivity allows for increased speed via 5G-NR radio access.

– 84% of cell users have access to 5G services thanks to the carrier aggregation technique. The control plane is carried out in the 700 MHz band.

5.2. Carrier aggregation

Carrier aggregation increases the bandwidth by combining several carriers. Each aggregated carrier is referred to as a component carrier (CC).

5G-NR CA supports up to 16 contiguous and non-contiguous CCs with different numerologies in the FR1 (Frequency Range 1) band and in the FR2 band.

The number of DC radio channels aggregated on the DL is greater than or equal to the number of DC radio channels aggregated on the UL uplink (UpLink). The numerology is associated with each radio channel and may be different from one radio channel to another.

Radio channels can be aggregated according to several patterns (Figure 5.2):

– radio channels may be contiguous in the same frequency band (intra-band contiguous CA);

– radio channels can be spaced out in the same frequency band with maximum frequency separation (non-contiguous intra-band CA);

– radio channels can be located in different frequency bands.

For the aggregation of contiguous intra-band carriers, the bandwidth and guard band are calculated from the configuration information of the edge DC radio channels $F_{edge,\ low}$ and $F_{edge,high}$.

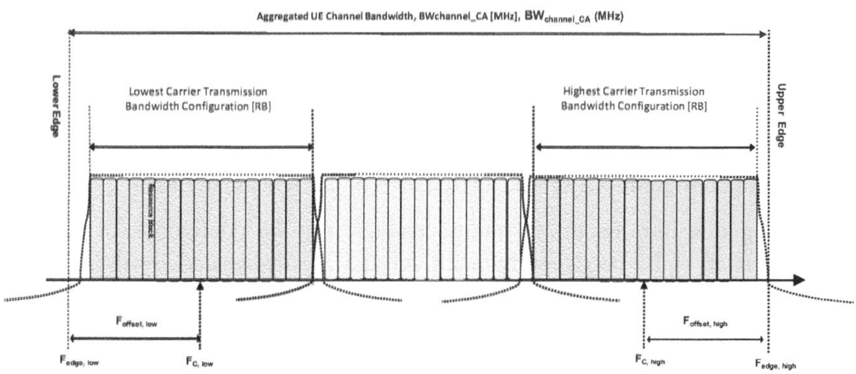

Figure 5.2. *The bandwidth of CA channels (extract from TS 38.101 Figure 5.3A.2-1)*

For the aggregation of non-contiguous intra-band carriers, the bandwidths are defined by sub-block; the sub-block may consist of the aggregation of several radio channels.

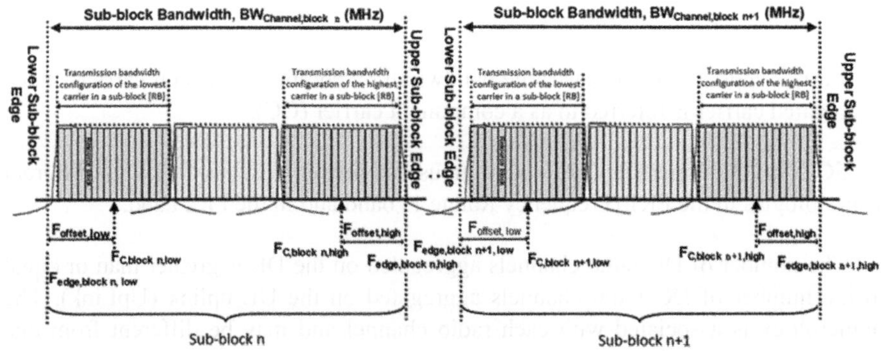

Figure 5.3. *Definition of the sub-block bandwidth for a non-contiguous intra-band spectrum (extract TS 38.101 Figure 5.3A.3-1)*

The radio channel aggregation notation determines the bandwidth class of the radio channel by the aggregation of multiple DC radio channels and the Fallback Group.

The class of radio bandwidth is associated with:

– a number of CC radio channels per sub-block;

– one bandwidth per radio channel;

– an aggregated channel bandwidth;

– a fallback group number.

When the terminal is configured with a bandwidth class, it may move to a lower class associated with the same fallback group or release a DC radio channel.

Different configurations for the aggregation of carriers are defined:

– Contiguous intra-band: the radio channels can be contiguous in the same frequency band (intra-band contiguous CA) and are designated by the following denomination CA_nX, Class, with nX being the number of the band and Class the category of the aggregation of carriers (example CA_n77C).

– Non-contiguous intra-band: the radio channels can be spaced in the same frequency band and are designated by the following denomination CA_nX (Class1,

Class2), with nX being the number of the band and Class the categories of the aggregation of carriers (example CA_n77 (3A–C)).

– Inter-band: the radio channels can be located in different frequency bands and are designated by the following denomination CA_nXA, nYA (example CA_n77-n79).

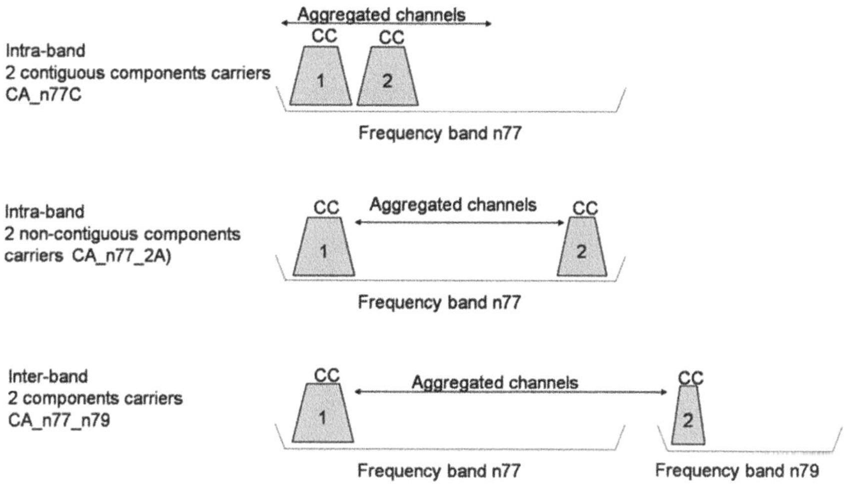

Figure 5.4. *Aggregation of component carriers*

The nX band numbers have been partially defined in Table 4.1 for the FR1 band and Table 4.2 for the FR2 band. Each band number is associated with a maximum bandwidth and the associated numerologies (see 3GPP TS 38.101-1 Table 5.3.5-1 for the FR1 band and 3GPP TS 38.101-2 Table 5.3.5-1 for the FR2 band).

The radio bandwidth classes are defined for each frequency band FR1 and FR2. In the case of the FR1 band, the inter-band configuration is limited to four bands. The inter-band configuration is not standardized in Release R.16 for the FR2 band.

5.2.1. *Carrier aggregation in the FR1 band*

5.2.1.1. *Carrier aggregation class*

In the FR1 band, for Release 16, the categories of radio bandwidth classes are defined in Table 5.1.

The maximum transmission band configuration is defined for each CC (Component Carrier) aggregate radio channel.

Class A corresponds to a 5G-NR configuration without carrier aggregation. The maximum 5G carrier band ($BW_{Channel,\ max}$) depends on the band number and the numerology. Numerology defines the frequency spacing SCS between sub-carrier (Sub Carrier Spacing). Class A belongs to all the fallback groups and allows the mobile to return to a basic configuration without aggregation of carriers.

Class B corresponds to a total bandwidth between 20 MHz and 100 MHz by the aggregation of two radio channels.

Class C corresponds to a total bandwidth between 100 MHz and 200 MHz from the aggregation of two radio channels.

Class D corresponds to a total bandwidth between 200 MHz and 300 MHz from the aggregation of three radio channels.

Class E corresponds to a total bandwidth between 300 MHz and 400 MHz from the aggregation of four radio channels.

Classes C, D and E belong to the same Fallback Group 1 fallback group.

Class G corresponds to a total bandwidth between 100 MHz and 150 MHz from the aggregation of three radio channels.

Class H corresponds to a total bandwidth between 150 MHz and 200 MHz from the aggregation of four radio channels.

Class I corresponds to a total bandwidth between 200 MHz and 250 MHz from the aggregation of five radio channels.

Class J corresponds to a total bandwidth between 250 MHz and 300 MHz from the aggregation of six radio channels.

Class K corresponds to a total bandwidth between 300 MHz and 350 MHz from the aggregation of seven radio channels.

Class L corresponds to a total bandwidth between 350 MHz and 400 MHz from the aggregation of eight radio channels.

Classes G to L belong to the same Fallback Group 2 fallback group.

The mobile can go from one class to another lower-level class belonging to the same fallback group, so, for example, the terminal can go from:

– class H to class G, from class G to class B or A;

– class D to class C, from class C to class A.

<table>
<tr><th>Aggregation class</th><th>Total bandwidth</th><th>Number of CC</th><th>Fallback group</th></tr>
<tr><td>A</td><td>$BW_{Channel} \leq BW_{Channel,max}$</td><td>1</td><td>1, 2</td></tr>
<tr><td>B</td><td>20 MHz $\leq BW_{Channel\_CA} \leq$ 100 MHz</td><td>2</td><td>2</td></tr>
<tr><td>C</td><td>100 MHz $< BW_{Channel\_CA} \leq 2 \times BW_{Channel,max}$</td><td>2</td><td rowspan="3">1</td></tr>
<tr><td>D</td><td>200 MHz $< BW_{Channel\_CA} \leq 3 \times BW_{Channel,max}$</td><td>3</td></tr>
<tr><td>E</td><td>300 MHz $< BW_{Channel\_CA} \leq 4 \times BW_{Channel,max}$</td><td>4</td></tr>
<tr><td>G</td><td>100 MHz $< BW_{Channel\_CA} \leq$ 150 MHz</td><td>3</td><td rowspan="6">2</td></tr>
<tr><td>H</td><td>150 MHz $< BW_{Channel\_CA} \leq$ 200 MHz</td><td>4</td></tr>
<tr><td>I</td><td>200 MHz $< BW_{Channel\_CA} \leq$ 250 MHz</td><td>5</td></tr>
<tr><td>J</td><td>250 MHz $< BW_{Channel\_CA} \leq$ 300 MHz</td><td>6</td></tr>
<tr><td>K</td><td>300 MHz $< BW_{Channel\_CA} \leq$ 350 MHz</td><td>7</td></tr>
<tr><td>L</td><td>350 MHz $< BW_{Channel\_CA} \leq$ 400 MHz</td><td>8</td></tr>
</table>

Table 5.1. *Carrier aggregation class on band FR1*

Carrier aggregation configuration includes:

– the type of carrier aggregation (intra-band, contiguous or not or inter-band);

– the number(s) of the 5G bands;

– the bandwidth class with the possible bandwidths on each CC.

5.2.1.2. *Contiguous intra-band carrier aggregation*

For the FR1 band, Release R.16 proposes 10 aggregation schemes of contiguous intra-band carriers (Table 5.2).

Table 5.3 lists non-exhaustive combinations of contiguous intra-band carrier aggregation for bands n7 and n77 and for classes B, C and D. Figure 5.4 illustrates the n77C configuration.

Band CA-NR	NR number
CA_n1	n1
CA_n7	n7
CA_n40	n40
CA_n41	n41
CA_n48	n48
CA_n66	n66
CA_n71	n71
CA_n77	n77
CA_n78	n78
CA_n79	n79

Table 5.2. *List of contiguous intra-bands for carrier aggregation FR1*

NR CA configuration	Bandwidth CC1 (MHz)	Bandwidth CC2 (MHz)	Bandwidth CC3 (MHz)	Cumulated maximal bandwidth (MHz)
CA_n7B	10, 15, 20	10, 15, 20, 30, 35, 40		50
CA_n77C	50	60, 80, 100		200
	60	60, 80, 100		
	80	80, 100		
	100	100		
	10, 15, 20, 25, 30, 40, 50, 60, 70, 80, 90, 100	10, 15, 20, 25, 30, 40, 50, 60, 70, 80, 90, 100		200
CA_n77D	100	100	100	300

Table 5.3. *Configuration for contiguous intra-band carrier aggregation*

5.2.1.3. *Non-contiguous intra-band carrier aggregation*

In the FR1 band, Release R.16 proposes eight non-contiguous intra-band carrier aggregation schemes (Table 5.4).

CA-NR number band	NR number band
CA_n3(*)	n3
CA_n7(*)	n7
CA_n25(*)	n25
CA_n41(*)	n41
CA_n48(*)	n48
CA_n66(*)	n66
CA_n77(*)	n77
CA_n78(*)	n78

Table 5.4. *Non-contiguous intra-band aggregation carrier in the FR1 band*

Table 5.5 lists non-exhaustive combinations of non-contiguous intra-band carrier aggregation for bands n7, n48 and n77.

Figure 5.4 illustrates the n77_2A configuration.

NR CA configuration	Bandwidth CC1 (MHz)	Bandwidth CC2 (MHz)	Bandwidth CC3 (MHz)	Bandwidth CC4 (MHz)	Cumulated bandwidth (MHz)
CA_n7(2A)	5, 10, 15, 20	5, 10, 15, 20			40
	10, 15, 20, 40, 50, 60, 80, 90, 100	10, 15, 20, 40, 50, 60, 80, 90, 100			140
CA_n48(3A)	10, 15, 20, 40, 50, 60, 80, 90, 100	10, 15, 20, 40, 50, 60, 80, 90, 100	10, 15, 20, 40, 50, 60, 80, 90, 100		140
CA_n48(4A)	10, 15, 20, 40, 50, 60, 80, 90, 100	10, 15, 20, 40, 50, 60, 80, 90, 100	10, 15, 20, 40, 50, 60, 80, 90, 100	10, 15, 20, 40, 50, 60, 80, 90, 100	135
CA_n77(2A)	20, 40, 80, 100	20, 40, 80, 100			200

Table 5.5. *Configuration of non-contiguous carrier aggregation intra-band*

The Release R.16 specification is limited aggregation class A configurations.

5.2.1.4. *Inter-band carrier aggregation*

The possible band combinations for the inter-band configuration are described in 3GPP TS 38.101-1 V16-4-0 specifications:

– 79 possible aggregations of two inter-band channels in the FR1 band (Table 5.2A.2.1-1);

– 35 possible aggregations of three inter-band channels in the FR1 band (Table 5.2A.2.2-1);

– six possible aggregations of four inter-band channels in the FR1 band (Table 5.2A.2.3-1).

Table 5.6 gives an example of aggregation on two bands, n77 and n79.

Figure 5.4 illustrates the n77A_n79A configuration.

NR CA configuration	NR band	SCS (kHz)	10 MHz	15 MHz	20 MHz	40 MHz	50 MHz	60 MHz	80 MHz	90 MHz	100 MHz
CA_n77A-n79A	n77	15	Yes	Yes	Yes	Yes	Yes				
		30	Yes	Yes	Yes	Yes	Yes	Yes	Yes	Yes	Yes
		60	Yes	Yes	Yes	Yes	Yes	Yes	Yes	Yes	Yes
	n79	15				Yes	Yes				
		30				Yes	Yes	Yes	Yes		Yes
		60				Yes	Yes	Yes	Yes		Yes
		30	Yes	Yes	Yes						
		60	Yes	Yes	Yes						

Table 5.6. *Inter-band carrier aggregation*

5.2.2. *Carrier aggregation in the FR2 band*

5.2.2.1. *Carrier aggregation class*

For the FR2 band and according to Release 16, the radio bandwidth class categories are defined in Table 5.7.

Class A corresponds to a 5G-NR configuration without carrier aggregation. The maximum 5G carrier band (BWChannel, max) depends on the band number and numerology. Class A belongs to all the fallback groups, thus allowing the mobile to return to a basic configuration without aggregation of carriers.

Class B corresponds to a total bandwidth between 400 MHz and 800 MHz from the aggregation of two radio channels.

Class C corresponds to a total bandwidth between 800 MHz and 1200 MHz from the aggregation of two radio channels.

Class B is a fallback configuration for class C, both of which belong to the same FallBack group 1 fallback list.

Class D corresponds to a total bandwidth between 200 MHz and 400 MHz from the aggregation of two radio channels.

Class E corresponds to a total bandwidth between 400 MHz and 600 MHz from the aggregation of three radio channels.

Class F corresponds to a total bandwidth between 600 MHz and 800 MHz from the aggregation of four radio channels.

Classes D, E and F belong to the same FallBack group 2 fallback list.

Class G corresponds to a total bandwidth between 100 MHz and 200 MHz from the aggregation of two radio channels.

Class H corresponds to a total bandwidth between 200 MHz and 300 MHz from the aggregation of three radio channels.

Class I corresponds to a total bandwidth between 300 MHz and 400 MHz from the aggregation of four radio channels.

Class J corresponds to a total bandwidth between 400 MHz and 500 MHz from the aggregation of five radio channels.

Class K corresponds to a total bandwidth between 500 MHz and 600 MHz from the aggregation of six radio channels.

Class L corresponds to a total bandwidth between 600 MHz and 700 MHz from the aggregation of seven radio channels.

Class M corresponds to a total bandwidth between 700 MHz and 800 MHz from the aggregation of eight radio channels.

Classes G, H, I, J, K, L and M belong to the same FallBack group 3 fallback list.

Class O corresponds to a total bandwidth between 100 MHz and 200 MHz from the aggregation of two radio channels.

Class P corresponds to a total bandwidth between 150 MHz and 300 MHz from the aggregation of three radio channels.

Class Q corresponds to a total bandwidth between 200 MHz and 400 MHz from the aggregation of four radio channels.

Classes O, P and Q belong to the same FallBack group 4 fallback list.

<table>
<tr><th>CA class</th><th>Cumulated bandwidth</th><th>Number of CC</th><th>Fallback group</th></tr>
<tr><td>A</td><td>$BW_{Channel} \leq 400$ MHz</td><td>1</td><td>1, 2, 3, 4</td></tr>
<tr><td>B</td><td>400 MHz $< BW_{Channel\_CA} \leq$ 800 MHz</td><td>2</td><td rowspan="2">1</td></tr>
<tr><td>C</td><td>800 MHz $< BW_{Channel\_CA} \leq$ 1200 MHz</td><td>3</td></tr>
<tr><td>D</td><td>200 MHz $< BW_{Channel\_CA} \leq$ 400 MHz</td><td>2</td><td rowspan="3">2</td></tr>
<tr><td>E</td><td>400 MHz $< BW_{Channel\_CA} \leq$ 600 MHz</td><td>3</td></tr>
<tr><td>F</td><td>600 MHz $< BW_{Channel\_CA} \leq$ 800 MHz</td><td>4</td></tr>
<tr><td>G</td><td>100 MHz $< BW_{Channel\_CA} \leq$ 200 MHz</td><td>2</td><td rowspan="7">3</td></tr>
<tr><td>H</td><td>200 MHz $< BW_{Channel\_CA} \leq$ 300 MHz</td><td>3</td></tr>
<tr><td>I</td><td>300 MHz $< BW_{Channel\_CA} \leq$ 400 MHz</td><td>4</td></tr>
<tr><td>J</td><td>400 MHz $< BW_{Channel\_CA} \leq$ 500 MHz</td><td>5</td></tr>
<tr><td>K</td><td>500 MHz $< BW_{Channel\_CA} \leq$ 600 MHz</td><td>6</td></tr>
<tr><td>L</td><td>600 MHz $< BW_{Channel\_CA} \leq$ 700 MHz</td><td>7</td></tr>
<tr><td>M</td><td>700 MHz $< BW_{Channel\_CA} \leq$ 800 MHz</td><td>8</td></tr>
<tr><td>O</td><td>100 MHz $\leq BW_{Channel\_CA} \leq$ 200 MHz</td><td>2</td><td rowspan="3">4</td></tr>
<tr><td>P</td><td>150 MHz $\leq BW_{Channel\_CA} \leq$ 300 MHz</td><td>3</td></tr>
<tr><td>Q</td><td>200 MHz $\leq BW_{Channel\_CA} \leq$ 400 MHz</td><td>4</td></tr>
</table>

Table 5.7. *Carrier aggregation of the FR2 band*

In case of fallback, the maximum bands of each aggregated band are:

– 100 MHz for FallBack group 4 fallback;

– 100 MHz for the FallBack group 3 fallback;

– 200 MHz for the FallBack group 2 fallback;

– 400 MHz for the FallBack group 1 fallback.

In the specification of Release R.16, only intra-band aggregations are standardized.

5.2.2.2. *Intra-band contiguous carrier aggregation*

In the FR2 band, Release R.16 proposes five aggregation schemes of contiguous intra-band carriers (Table 5.8).

NR CA band	NR band
CA_n257	n257
CA_n258	n258
CA_n259	n259
CA_n260	n260
CA_n261	n261

Table 5.8. *Carrier aggregation intra-band in FR2*

Table 5.9 lists non-exhaustive combinations of contiguous intra-band carrier aggregation for the n257 bands for different classes B, C, D, E and F.

The list of possible configurations is described in Table 5.5A.1.1 of specification TS 38.101-2.

NR CA configuration	Bandwidth CC1 (MHz)	Bandwidth CC2 (MHz)	Bandwidth CC3 (MHz)	Bandwidth CC4 (MHz)	Cumulated bandwidth (MHz)
CA_n257B	50, 100, 200, 400	400			800
CA_n257C	50, 100, 200, 400	400	400		1200
CA_n257D	50, 100, 200	200			400
CA_n257E	50, 100, 200	200	200		600
CA_n257F	50, 100, 200	200	200	200	800

Table 5.9. *Contiguous intra-band carrier aggregation in the FR2 band*

5.2.2.3. *Non-contiguous intra-band aggregation*

For the FR2 band, Release R.16 provides a list of possible configurations for downlink:

– in Table 5.5A.2.1 of specification TS 38.101-2 for configurations of the same configuration class such as CA_n260 (2A) or CA_n260 (2D);

– in Table 5.5A.2.2 of specification TS 38.101-2 for configurations of the same configuration class such as CA_n260 (A-D) or CA_n260 (5A-2O).

In the case of aggregation of non-contiguous carriers, the frequency spacing between each CC is normalized by the separation class shown in Table 5.10.

Frequency separation class	Maximum authorized frequency separation (Fs)
I	800 MHz
II	1200 MHz
III	1400 MHz
IV	1000 MHz
V	1600 MHz
VI	1800 MHz
VII	2000 MHz
VIII	2200 MHz
IX	2400 MHz

Table 5.10. *Maximum spacing of CC radio channels*

5.3. Supplementary UpLink (SUL)

Carrier aggregation increases radio data rates by adding the bandwidth of each CC radio channel.

In the case of SUL mode, the mobile terminal has an additional uplink band, but uplink transmission can only take place in one band, either the traditional paired band or the SUL band.

The SUL bandwidth is generally smaller than the traditional band, but the carrier frequency is lower, thereby improving cell coverage for the uplink.

The SUL bandwidth can be automatically adapted by the radio node if the latter manages the DSS mode with a sharing of the 4G/5G band.

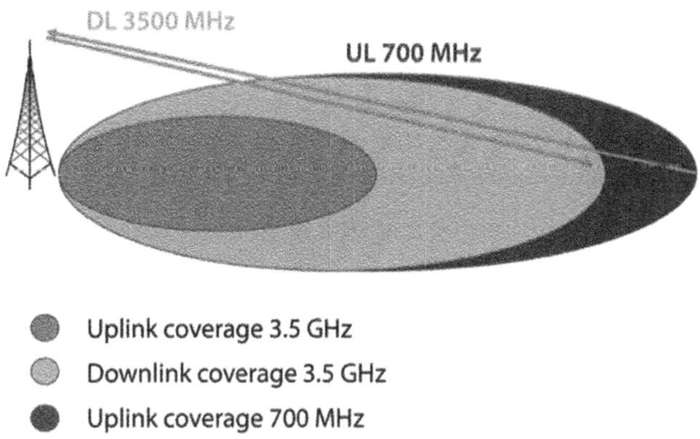

Figure 5.5. *SUL mode. For a color version of this figure, see www.iste.co.uk/launay/5g.zip*

5.4. Synchronization on the secondary cell

5.4.1. *Carrier aggregation procedure*

5.4.1.1. *The activation and deactivation of the secondary cell*

Connection establishment is triggered by the mobile to the primary cell (PCell).

The PCell supports:

– random access and RRC (Radio Resource Control) connection mechanisms;

– NAS (Non-Access Stratum) messages relating to mobility management and session management.

The other radio channels are called SCell (Secondary Cell). The secondary cells only transmit IP packets of the traffic plane.

The aggregation can associate radio channels operating in TDD and FDD modes depending on the capabilities of the mobile UE regarding:

– the combination list of the aggregated bands *BandCombinationList*;

– the bandwidth classes of the *CA-BandwidthClassNR* carrier aggregation;

– *CA-ParameterNR* carrier aggregation parameters.

The parameters concerning the aggregation of carriers allow us to specify whether the mobile supports additional or optional features such as:

– the configuration of a TAG (Timing Advance Group) for each cell, required in particular in the case of heterogeneous networks for which the secondary cell is carried by a remote radio node RRH (Radio Remote Head);

– simultaneous transmission of the SRS reference signal and the PUCCH and PUSCH physical channels on different CC radio channels (parallelTxSRS-PUCCH-PUSCH);

– simultaneous transmission of the PRACH random access channel and SRS/PUCCH/PUSCH signals through different DC radio channels (parallelTxSRS-PUCCH-PUSCH);

– simultaneous TDD-TDD and TDD-FDD inter-band communication (simultaneousRxTxInterBandCA);

– the use of different numerologies on each CC radio channel (diffNumerologyAcrossPUCCH-Group).

The carrier aggregation procedure is established by the RRC control message.

The activation and deactivation of the SCell secondary cells is carried out by control messages on layer 1 (physical) and layer 2 (MAC) called L1/L2 Control Signaling and carried by the PDCCH/PDSCH channel (DCI information: Downlink Control Information).

The DCI information is transmitted either only on the primary cell PCell (Cross Carrier Scheduling) or on each CC PCell and SCell (Self Carrier Scheduling):

1) the gNB entity sends the RRC *Reconfiguration* message to the mobile indicating the configuration of the radio resource;

2) the mobile applies the new configuration and responds to the gNB entity with the RRC *ReconfigurationComplete* message;

3) secondary cells can be quickly turned on or off through the MAC CE (Medium Access Control Control Element) layer.

The activation and deactivation of the cell is carried out by the LCID index 111001 followed by a 4-byte message or the LCID index 111010 followed by a one-byte message.

The LCID (Logical Channel Identifier) is performed at the MAC layer to reassemble PDCP frames from the same medium.

The message enables or disables a secondary cell by respectively setting the bit to 1 or 0.

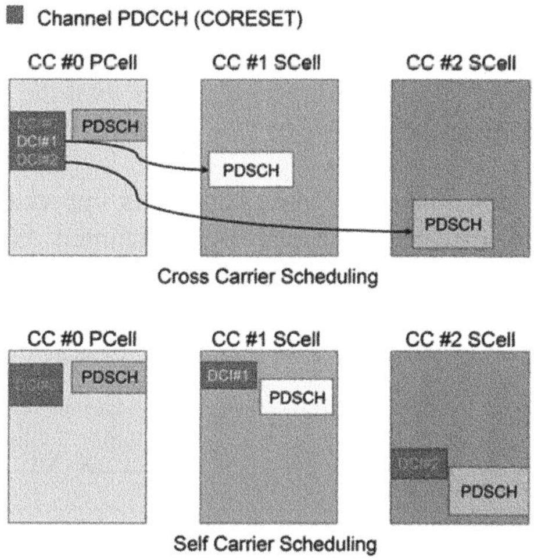

Figure 5.6. *L1/L2 control signaling. For a color version of this figure, see www.iste.co.uk/launay/5g.zip*

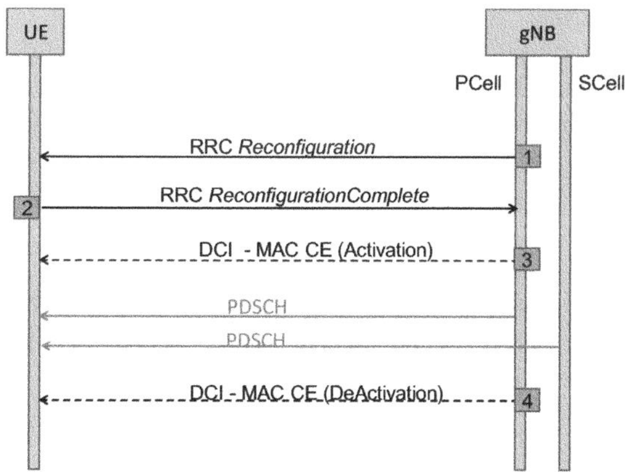

Figure 5.7. *Carrier aggregation procedure. For a color version of this figure, see www.iste.co.uk/launay/5g.zip*

5.4.1.2. *HARQ procedure*

HI (HARQ Indicator) information relates to the positive ACK, or negative NACK, acknowledgment of data received on the physical PDSCH channel.

HI information is transmitted in the PUCCH physical channel according to one of the UCI (Uplink Control Information) formats defined in Chapter 4.

In the case of carrier aggregation, the UCI control information is transmitted in the PUCCH physical channel of the primary cell PCell so as to allow a number of DC radio channels in the downlink direction greater than the number of DC radio channels in the uplink direction.

If the number of downlink CC radio channels is important, then the PUCCH channel on the uplink can be simultaneously transmitted on two radio channels PCell and SCell to avoid saturation of uplink PUCCH in the PCELL.

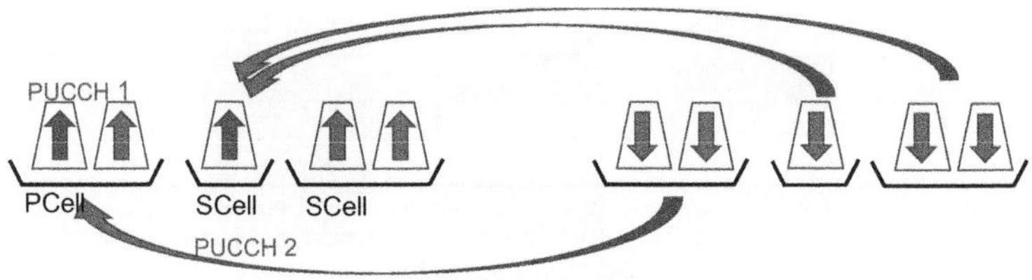

Figure 5.8. *The transmission of acknowledgments on two PUCCH channels. For a color version of this figure, see www.iste.co.uk/launay/5g.zip*

5.4.2. *SUL procedure*

The SUL procedure consists of adding an additional radio channel for uplink direction in order to improve the uplink coverage.

The SUL radio channel is on a lower carrier frequency and of lower bandwidth compared to the traditional (non-SUL) channel.

The SUL channel can be a shared channel with an LTE frequency band, in the case of NR/LTE coexistence and whose spectrum is dynamically allocated by the base station (Dynamic Spectrum Sharing).

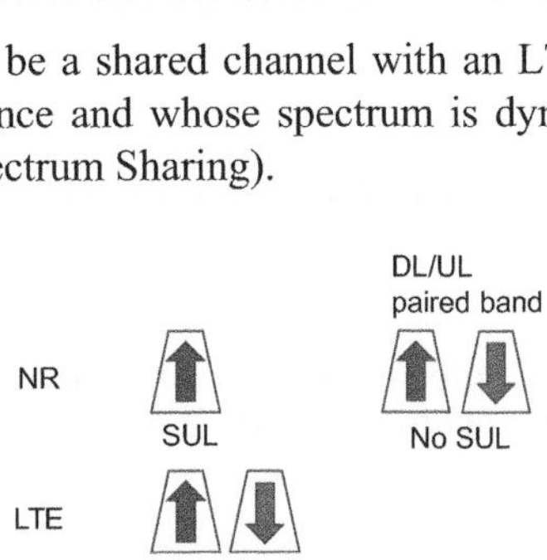

Figure 5.9. *SUL mode with coexistence LTE and without coexistence. For a color version of this figure, see www.iste.co.uk/launay/5g.zip*

The uplink radio channel (SUL or not SUL) is controlled by the radio node to avoid transmission of the PUSCH physical channel or the PUCCH physical channel on both SUL and non-SUL radio channels simultaneously.

The random access procedure (see Chapter 4), allowing the radio connection between the mobile and the base station, can be performed on any of the two bands.

The selection of the SUL or non-SUL channel for the initial access is imposed by the radio node or selected by the mobile by comparing the power of the downlink signal to a power threshold. The sul-RSRP-Threshold power threshold value is transmitted by the broadcast signal SIB1 (Signal Information Block) in the *servingCellConfigCommon* data structure.

The SUL channel can supplement the carrier aggregation technique. It is possible to add an SUL channel on the primary PCell or secondary SCell.

5.5. References

All standards can be downloaded from the ETSI website: https://www.etsi.org/standards.

3GPP TR 38.716-01-01
Technical Specification Group Radio Access Networks;
NR intra band Carrier Aggregation (CA) Rel-16 for xCC Down Link (DL)/yCC Up Link (UL) including contiguous and non-contiguous spectrum (x >= y)
Version 16.0.0 Release 16

3GPP TS 38.101-01
NR - User Equipment (UE) radio transmission and reception;
Part 1: Range 1 Standalone
Version 16.4.0 Release 16

3GPP TS 38.101-02
NR - User Equipment (UE) radio transmission and reception;
Part 2: Range 2 Standalone
Version 16.4.0 Release 16

3GPP TS 38.104
NR - Base Station (BS) radio transmission and reception;
Version 16.4.0 Release 16

3GPP TR 38.133
NR - Requirements for support of radio resource management;
Version 16.4.0 Release 16

3GPP TR 38.321
NR - Medium Access Control (MAC);
Version 16.4.0 Release 16

3GPP TR 38.331
NR - Radio Resource Control (RRC);
Version 15.2.1 Release 15

Ericsson (2020). Carrier aggregation in 5G [Online]. Available at: https://www.ericsson.com/en/networks/offerings/5g/carrier-aggregation.

6

5G-NR Radio Interface – MIMO and Beamforming

6.1. Multiplexing techniques

6.1.1. *MIMO mechanism*

The MIMO mechanism improves radio data rate by performing spatial multiplexing of multiple signals. Each signal is transmitted on several antennas and received on multiple antennas, and all signals share the same time–frequency resource.

SU-MIMO (Single User – Multiple Input Multiple Output) refers to the transmission from a base station to one mobile, using the same frequency and time resource, on two antennas (2x2 MIMO) (Figure 6.1), on four antennas (4x4 MIMO) or on eight antennas (8x8 MIMO).

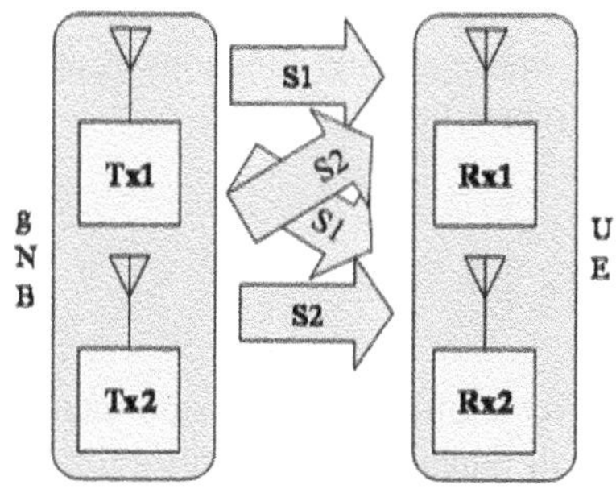

Figure 6.1. *SU-MIMO mechanism*

For a color version of all the figures in this chapter, see www.iste.co.uk/launay/5g.zip.

MU-MIMO (Multi User – MIMO) refers to the transmission from a radio node to different mobiles, using the same frequency and time resource, on two antennas (2x2 MIMO) (Figure 6.2), on four antennas (4x4 MIMO) or on eight antennas (8x8 MIMO).

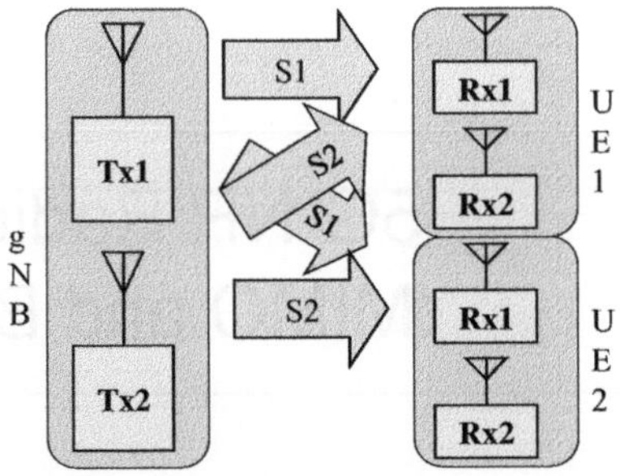

Figure 6.2. *MU-MIMO mechanism*

The receivers receive the components y_1 and y_2 which are the product of the components x_1 and respectively x_2 (the signals transmitted) by the transmission matrix H. The signal x_1 (respectively x_2) is transmitted by the transmitter Tx1 (respectively Tx2). The signal y_1 (respectively y_2) is received by the receiver Rx1 (respectively Rx2). The transmission matrix H contains the transfer functions h_{ij}, from the transmitter j to the receiver i.

$$\begin{bmatrix} y_1 \\ y_2 \end{bmatrix} = \begin{bmatrix} h_{11} h_{12} \\ h_{21} h_{22} \end{bmatrix} \times \begin{bmatrix} x_1 \\ x_2 \end{bmatrix}$$

Spatial demultiplexing consists of recovering the components x_1 and x_2 of the transmitted signal from the components y_1 and y_2 of the received signal, and from the transmission matrix H, whose transfer functions h_{ij} are calculated from reference signals transmitted by the transmitter.

$$\begin{bmatrix} x_1 \\ x_2 \end{bmatrix} = \begin{bmatrix} h_{11} h_{12} \\ h_{21} h_{22} \end{bmatrix}^{-1} \times \begin{bmatrix} y_1 \\ y_2 \end{bmatrix}$$

The estimate of the inverse of the transmission matrix is usually calculated by the MMSE (Minimum Mean Square Error) equalization method or by the ZF (Zero Forcing) equalization method.

6.1.2. *Baseband beamforming*

Beamforming is a complementary mechanism to MIMO.

Beamforming uses multiple antennas to control the direction of the beam by individually weighting the amplitude and phase of each transmitted signal, by applying a specific precoding matrix to each component of the transmitted signal (Figure 6.3) in baseband.

The precoding vector is determined by the mobile from the reference signals and transmitted to the base station. Beamforming logically reduces the beam opening angle and limits the level of interference between cells. The beamforming allows us to increase the range thanks to the increase in power as a result of the contribution of each transmitted signal.

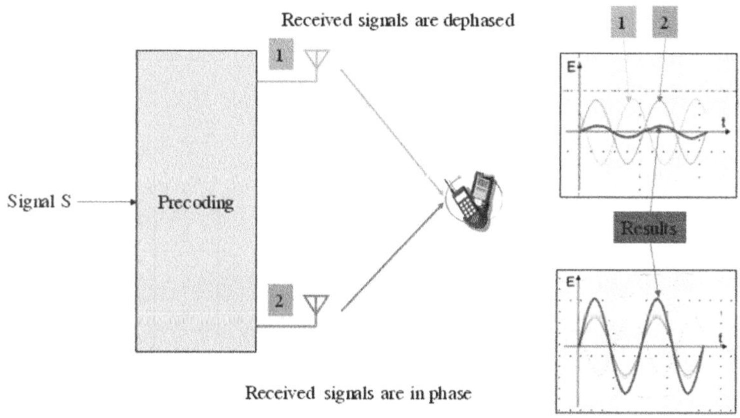

Figure 6.3. *Beamforming*

6.1.3. *Active antennas and massive-MIMO*

Elevation Beamforming (EBF) allows a beam to be directed in a specific way, for example, a beam pointing either up or down. This technique contrasts with conventional antenna systems for which the tilt angle is fixed.

The Active Antenna System (AAS) mechanism is based on the formation of elevation beams. The horizontal and vertical directions can be combined, leading to two-dimensional beam formation (Figure 6.4).

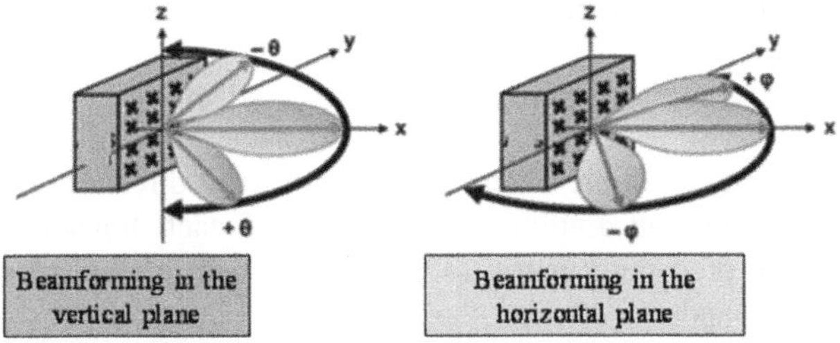

Figure 6.4. *Beamforming in horizontal and vertical planes (source: NTT Docomo Technical Journal Vol. 18, No. 2)*

This new beamforming technique tends to increase the number of antennas as the signal has spatial access to more degrees of freedom.

The active antenna system (AAS) consists of a block of TXU transmitters and TRU receivers, an RDN (Radio Distribution Network) and an AA (Antenna Array) network (Figure 6.5).

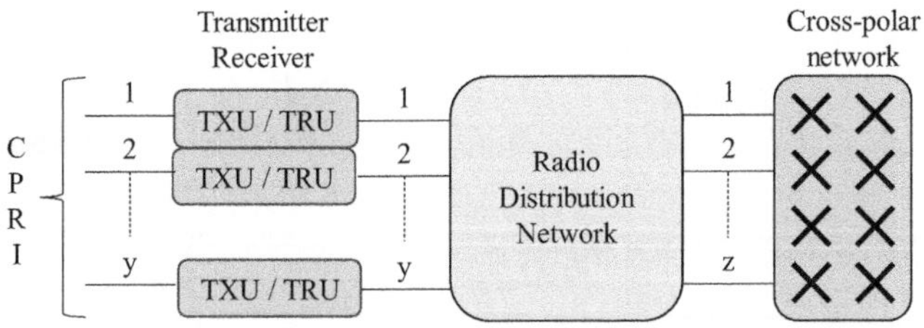

Figure 6.5. *Active antennas*

The TXU transmitter converts baseband signals from the Common Public Radio Interface (CPRI) into radio frequency signals. Each baseband signal can be assigned to one or more TXU transmitters. The radio frequency outputs are distributed to the array of antennas via the radio distribution network. The TRU receiver converts RF signal to baseband.

In order to provide an elevation of zenith angle θ_{etilt}, each radio frequency signal supplied by the TXU unit is weighted by a complex coefficient and distributed to different radiated elements by the RDN.

The same operation is performed for the signal received from the antenna network.

The configuration of the antenna array is given by the parameters (M, N and P), where M is the number of antenna elements having the same polarization in each column (M = {1, 2, 4, 8 }), N is the number of columns (N = {1, 2, 4, 8, 16}) and P is the number of polarizations (normally equal to 2).

M_{TXRU} is the number of TXU transmitters and TRU receivers per column and per polarization and N_{TXRU} is the number of TXU/TRUs per row and per polarization.

Two generic mapping models between, on the one hand, antenna elements {M, N, P} and, on the other hand, TXU transmitters and TRU receivers {M_{TXRU}, N_{TXRU}} have been defined; the antenna partitioning model and the full connection model.

For the antenna partitioning model (Figure 6.6), the antenna elements are distributed among different groups, and each TXU/TRU is connected to a group. The value of the weighting w_k of the radio signal is given by the following formula:

$$w_k = \frac{1}{\sqrt{K}} \exp\left(-j\frac{2\pi}{\lambda}(k-1)d_v \cos\theta_{etilt} \right), \text{ for } k=1, 2, \ldots, K$$

where:

$$K = \frac{M}{M_{TXRU}};$$

λ is the wavelength;

d_v is the distance between two vertical antenna elements;

θ_{etilt} is the zenith angle.

For the full connection model (Figure 6.6), each TXU/TRU is connected to all antenna elements. A combiner is used to connect the different TXU/TRU units to each antenna element. The value of the weighting $w_{m,m'}$ of the radio signal is given by the following formula:

$$w_{m,m'} = \frac{1}{\sqrt{M}} \exp\left(-j\frac{2\pi}{\lambda}(m-1)d_v \cos\theta_{etilt,m'} \right)$$

for $m=1, 2, \ldots, M$ and $m'=1, 2, \ldots, M_{TXRU}$

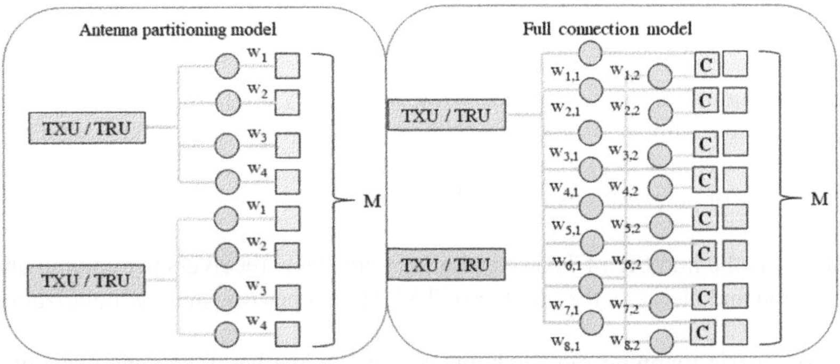

Figure 6.6. *Mapping models between TXU/TRU units and the antenna elements*

The weighting vector is implemented at the radio frequency signal level to beam the signal in a given direction.

The Observed Time Difference Of Arrival (OTDOA) angle estimation method measures the signal reception power on each radiating element. ESPRIT (Estimation of Signal Parameters via Rotational Invariance) or MUSIC (Multi-Signal Classification) algorithms are the most used to define the angle of arrival. The advantage of this method is that it does not require information about the transmission channel.

Active antenna systems control the beam in an analog (RF) way. 8 to 16 TXU/TRU radio frequency channels are connected to 16–32 radiating elements.

The massive-MIMO mechanism corresponds to the increase in the number of M antennas, far greater than the number of K antennas of a UE mobile: 32, 64 and 128 TXU/TRU radio frequency chains are connected to 128, 192 and 256 radiating elements.

The massive-MIMO 32T32R, 64T64R and 128T128R include 32, 64 and 128 RF channels (TXU/TRU), respectively.

The massive-MIMO makes it possible to combine beam generation by digital coding and improves coverage by a connecting network of analog antennas.

The hybrid architecture makes it possible to couple the techniques of baseband multiplexing (MIMO, beamforming) with an analog beam orientation to provide antenna gain (better coverage).

The digital architecture is simpler to implement in the FR1 band.

The analog architecture is optional in the FR1 band; however, it is mandatory in the FR2 band.

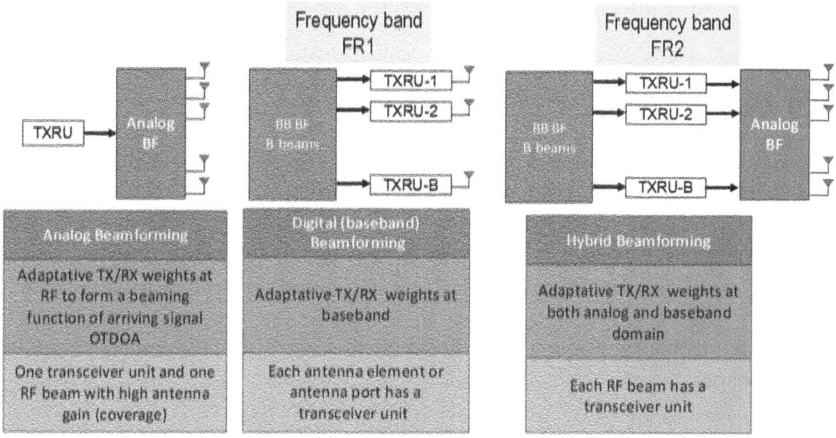

Figure 6.7. *Beamforming architecture*

The main evolutions of the massive-MIMO mechanism compared to the traditional MIMO mechanism are provided in Table 6.1.

Mechanism	MIMO	AAS	massive-MIMO
Number of antenna ports CSI-RS	1, 2, 4, 8	1, 2, 4, 8, 12, 16	1, 2, 4, 8, 12, 16, 20, 24, 28, 32
Number of RF chains	1–8	8–16	32–128
Number of antennas	1–8	16–32	128–256
Beam control	Baseband	RF	Hybrid
Beamforming	Horizontal plane	Vertical and horizontal plane	3D beam
Number of RF signals SU-MIMO	8	8	8
Number of RF signals MU-MIMO	4 (note 1)	8 (note 2)	8 (note 2)

Note 1: the maximum number of spatially multiplexed mobiles is equal to 4. The number of radio signals per mobile is then equal to 2.

Note 2: the maximum number of spatially multiplexed mobiles is equal to 8. The maximum number of radio signals per mobile is equal to 2.

Table 6.1. *Comparison between MIMO and massive-MIMO mechanisms*

6.1.4. *Antenna systems*

The transmission modes implementing beamforming and the MU-MIMO mechanism require good correlation between the antennas.

The transmission modes implementing transmission diversity and the SU-MIMO mechanism require decorrelation between the antennas.

For an antenna made up of columns of vertically polarized radiating elements, the correlation between the columns is relatively strong.

For an antenna consisting of a column of two sets of radiating elements, each set corresponding to ±45 degree cross-polarization, the correlation is relatively weak.

The antenna configurations are described in Figure 6.8 and relate to a single frequency band.

Configuration A corresponds to an antenna made up of a column of vertically polarized radiating elements.

Configuration B corresponds to an antenna made up of two columns of vertically polarized radiating elements.

Configuration C corresponds to an antenna made up of four columns of vertically polarized radiating elements.

Configuration D corresponds to an antenna made up of a column of two sets of radiating elements, each set corresponding to a cross-polarization ±45 degrees.

Configuration E corresponds to an antenna consisting of two columns, each column comprising two sets of radiating elements and each set corresponding to a cross-polarization ±45 degrees.

Configuration E combines pairs of correlated radiating elements of the same polarization and pairs of decorrelated radiating elements of different polarizations.

Configuration F corresponds to an antenna consisting of:

– a central column of two sets of radiating elements, each set corresponding to a cross-polarization ±45 degrees;

– two sets which are distant spatially, each set comprising radiating elements, corresponding to one of the two cross-polarizations ±45 degrees.

The remaining polarization is available for use in another frequency band.

Configuration G corresponds to an antenna made up of four columns of radiating elements, each column comprising two sets of radiating elements and each set corresponding to a cross polarization ±45 degrees.

Configuration H corresponds to two spaced antennas, each antenna consisting of a column of two sets of radiating elements and each set corresponding to a cross polarization ±45 degrees.

Configuration I is two spaced antennas, each antenna consisting of a column of vertically polarized radiating elements.

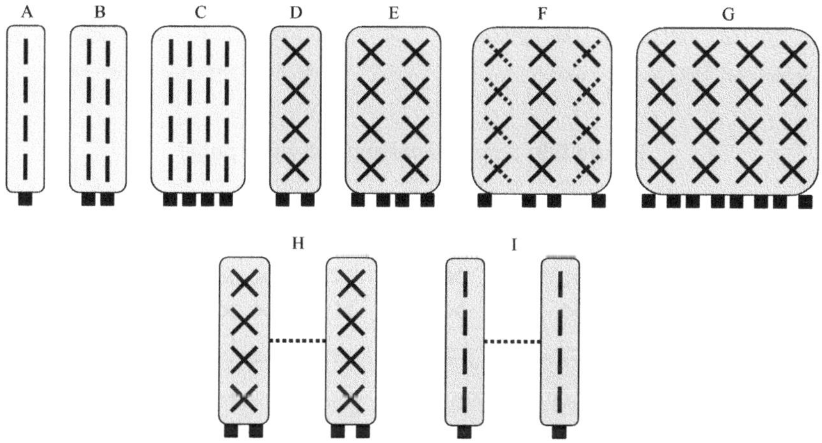

Figure 6.8. *Antenna configuration*

The antenna structure is formed of a panel (single-panel) or a panel assembly (multi-panel).

The single-panel structure is based on the full connection model: the RF chains are connected to all the antenna elements.

The multi-panel structure consists of connecting several RF chains to different panels. The RF chains are grouped into a sub-assembly, and each sub-assembly is connected to a single-panel structure.

Antenna structure is defined by three parameters (Ng, N1 and N2), with Ng being the number of panels, N1 being the number of columns per panel and N2 being the number of rows per panel.

In Figure 6.9, the antenna structure is (1,8,2) on the left and (2,4,2) on the right.

In Figure 6.10, the antenna structure is (2,8,2).

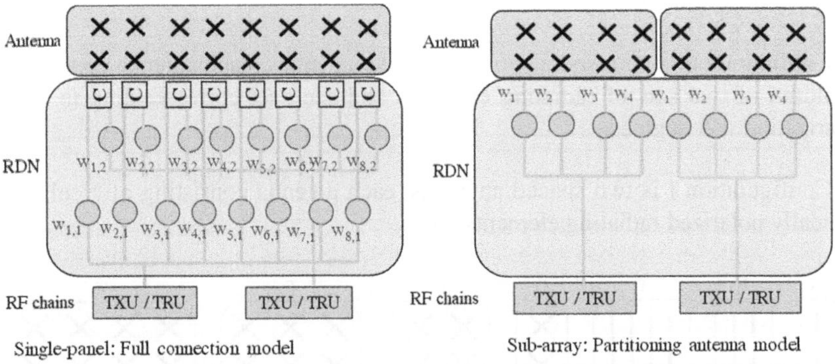

Figure 6.9. *Single panel and sub-array panel*

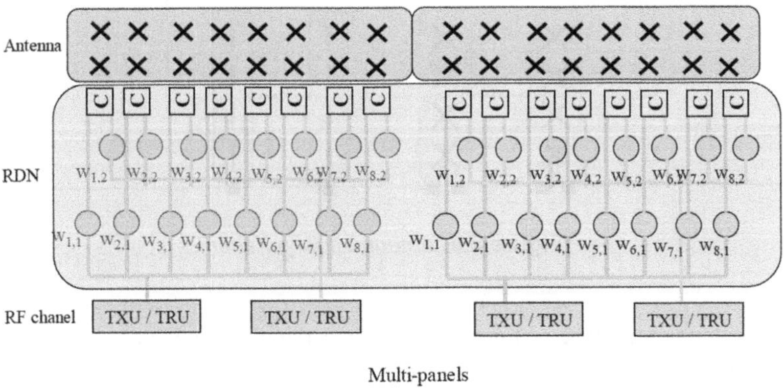

Figure 6.10. *The structure of multi-panels*

6.2. Antenna port

The MIMO mechanism and the beamforming mechanism are based on knowledge of the propagation channel.

On the uplink (UL), the mobile uses the reference signals sent by the base station to estimate the propagation channel.

On the downlink (DL), the base station uses the following:

– the reference signals transmitted by the mobile;

– and/or the measurement report made by the mobile from the reference signals sent by the base station (feedback method).

The CSI (Channel State Information) measurement report is transmitted from the mobile to the base station via a control information message on the UCI (Uplink Control Information) uplink.

The CSI-RS reference signals for the downlink or SRS for the uplink channel are used for the equalization of channel Matrix H.

The particularity of the TDD duplexing mode in 5G is the ability to use reference signals transmitted by the mobile.

The concept of antenna port refers to a logical antenna carrying a reference signal associated or not with physical channels.

The antenna port can correspond to several physical antennas, and the receiver estimates the propagation channel between the logical transmitting antenna and the receiving antenna.

Knowing the reference signal, the receiver reconstructs the transfer function of the propagation channel from the distortion of the received signal. This estimate is theoretically valid only on the tempo-frequency resources to which the reference signal is mapped. However, due to the stationarity in time and frequency, the estimate remains valid for the duration of a slot, on a block resource for an antenna port.

When two signals are transmitted on different antenna ports, the propagation channel differs; however, in the case of the quasi-stationary channel, the signal distortion is similar from one antenna port to another. Thus, the estimation of the propagation channel of the signal emitted by one antenna can be used for the estimation of the propagation channel of the signal emitted on the other antenna. This technique is called QCL (Quasi-CoLocation) mode.

6.2.1. *Downlink transmission*

Symbols produced by modulation are distributed over spatial layers, and then precoded before being assigned to resource elements by antenna port.

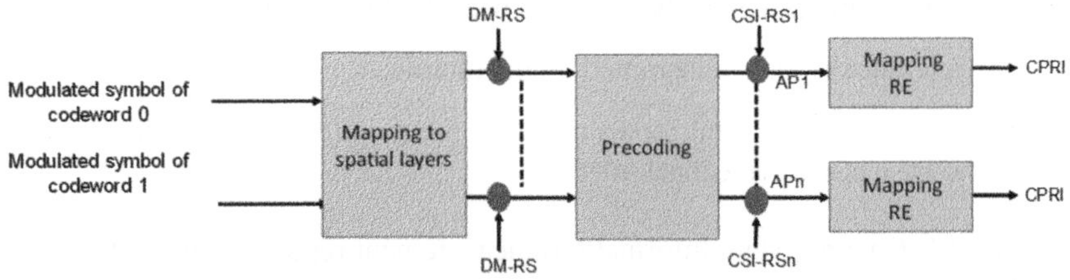

Figure 6.11. *MIMO transmitter scheme*

The precoded symbols are associated with reference signals (RS) to form antenna ports.

Table 6.2 shows the association of the antenna ports and the downlink reference signals.

The DM-RS associated with the PDSCH can be transmitted over eight spatial layers on antenna ports from 1000 to 1007 for the DM-RS type 1 configuration and from 1000 to 1011 for the DM-RS type 2 configuration (see Chapter 4).

The PT-RS associated with the PDSCH is a repetition of the DM-RS on antenna ports 1000 to 1005.

The DM-RS associated with the PDCCH is only transmitted over a single 2000 antenna port.

The Synchronization Signal Block (SSB) is transmitted on only one antenna port 4000.

Antenna port	Reference signal
P1000–P1011	DM-RS (PDSCH)
P1000–P1005	PT-RS (PDSCH)
P2000	DM-RS (PDCCH)
P3000–P3031	CSI-RS
P4000	Block SSB
P5000	PRS

Table 6.2. *Mapping of antenna ports and reference signals in downlink transmission*

6.2.1.1. *The CSI-RS physical signal*

The CSI-RS physical signal (Channel State Information Reference Signal) is used to perform coherent demodulation of the received signal.

The coherent demodulation of the received signal is based on the estimation of the transfer function of the radio channel.

The CSI-RS physical signal allows the implementation of spatial multiplexing and transmission diversity.

The strength of the CSI-RS physical signal is either transmitted to determine the level of the received signal or removed to measure the level of interference.

The CSI-RS physical signal is used to measure the RSRP (Reference Signal Received Power) and the RSRQ (Reference Signal Received Quality).

6.2.1.2. *The MBSFN-RS physical signal*

The MBSFN (MBMS Single Frequency Network) RS physical signal is transmitted only in the PMCH (Physical Multicast Channel) to perform coherent demodulation of the received signal.

The PMCH physical channel is used to transmit IP (Internet Protocol) packets in broadcast mode.

6.2.1.3. *The DM-RS physical signal*

The DM-RS physical signal is used for the beamforming mechanism and for spatial multiplexing of several users.

The DM-RS signal is associated with a PDSCH (Physical Downlink Shared Channel), a PDCCH (Physical Downlink Control Channel) or a PBCH (Physical Broadcast Channel) to allow demodulation of the associated physical channel.

The DM-RS physical signal allows the following configurations:

– one user using a single spatial layer;

– one user, using two spatial layers, and a beamforming associated with MIMO 2x2 spatial multiplexing;

– two users, using two spatial layers, the multiplexing of the two users being carried out by an identity code;

– four users, using a spatial layer, the multiplexing of the four spatial layers being obtained from an OCC (Orthogonal Covering Code) and an identity code;

– a user using eight spatial layers, and a beamforming associated with 8x8 MIMO spatial multiplexing.

6.2.1.4. *The PRS physical signal*

The PRS (Positioning Reference Signal) physical signal is used by the mobile for the implementation of the OTDOA (Observed Time Difference Of Arrival) function.

The OTDOA function is based on the measurement by the mobile of the difference in the reception time of the physical PRS signal compared to a reference cell.

The location of the mobile is obtained from three measurements taken on three geographically dispersed cells.

6.2.2. *Uplink transmission*

Table 6.3 shows the association of antenna ports and uplink channels and physical signals.

Antenna port	Reference signal
P0–P7	DM-RS (PUSCH)
P1000	SRS
P2000	DM-RS (PUCCH)
P4000	PRACH

Table 6.3. *Numbering of the antenna port in uplink transmission*

6.2.2.1. *The SRS physical signal*

The SRS (Sounding Reference Signal) physical signal is used to perform coherent demodulation of the received signal.

The coherent demodulation of the received signal is based on the estimation of the transfer function of the radio channel.

The SRS physical signal allows the implementation of spatial multiplexing and transmission diversity.

6.2.2.2. *The DM-RS physical signal in uplink*

The DM-RS physical signal is used for the beamforming mechanism and for spatial multiplexing of several users. The DM-RS allows the demodulation of the physical channel with which it is associated, either the PUCCH (Physical Uplink Control Channel) or the PUSCH (Physical Uplink Shared Channel).

The DM-RS physical signal on the uplink allows a user to exploit up to four spatial layers.

6.2.2.3. *The PRACH physical signal*

The PRACH (Physical Random Access Channel) signal carries the random access preamble.

6.3. Uplink Control Information (UCI)

UCI represents the data transmitted by the mobile which contains the scheduling requests (*SR*), the HI hybrid acknowledgment (HARQ Indicator) and the CSI report.

The HI information concerns the positive ACK or negative NACK acknowledgment of data received on the PDSCH physical channel.

The CSI, determined by the UE, groups together information relating to the status reports of the signal received on the PDSCH physical channel and received at the SSB block:

– the CQI (Channel Quality Indicator) information corresponds to the modulation and coding scheme recommended for the PDSCH physical channel;

– the RI (Rank Indicator) information determines the number of spatial layers recommended for the PDSCH physical channel;

– the PMI (Precoder Matrix Indicator) information provides indications relating to the precoding matrix in the case of use of spatial multiplexing, in closed loop, or in the case of beamforming.

The reporting of CSI can be periodic or aperiodic:

The aperiodic report is always transferred to the PUSCH (Physical Uplink Shared Channel) physical channel.

The periodic report is transferred to the PUCCH (Physical Uplink Control Channel) in the following two cases:

– no resource is allocated to the mobile in the PUSCH physical channel;

– a resource is allocated to the mobile, and simultaneous transmission of the PUCCH and PUSCH physical channels is possible.

Otherwise, the periodic report is transmitted in the PUSCH physical channel.

The report transfer modes are built according to the type of feedback from the CQI and PMI information.

The type of transfer of the aperiodic or periodic report and the types of measurements are communicated to the mobile by the RRC Setup or RRC Reconfiguration messages.

The aperiodic report is used to report CSI relating to the BWP (Bandwidth Partition) of the radio channel or to sub-bands of the radio channel, depending on the format indicator *cqi-FormatIndicator*.

The transfer of the aperiodic report is triggered by the following messages:

– DCI (Downlink Control Information) transmitted in the PDCCH (Physical Downlink Control Channel);

– the RAR (Random Access Response) message transmitted in the PDSCH.

6.4. PDSCH transmission

6.4.1. *Single-CSI and multiple-CSI transmission*

Whereas 10 transmission modes (TM) have been defined for 4G, there is only one transmission mode in Release R.16 for the PDSCH. Nevertheless, this transmission mode makes it possible to implement the MIMO mechanism using a baseband precoder. The precoder is defined by its matrix W, and the baseband beamforming mechanism is defined by a beam precoding X matrix.

All the codes are registered in a codebook. The base station chooses the precoding code from the CSI measurement report sent by the mobile. The mobile calculated CSI from the CSI-RSs.

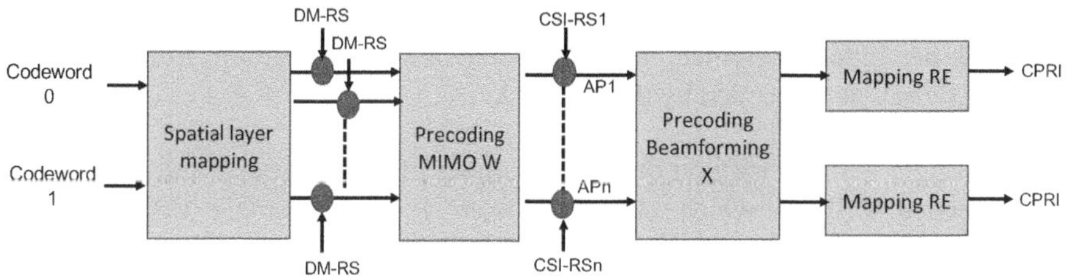

Figure 6.12. *Massive-MIMO transmission chain*

The CSI-RS are not precoded by the precoding matrix W but can be precoded by beamforming X (see Figure 6.12). The received CSI-RS signal corresponds to the convolution product [H].[X].

The DM-RS are precoded (see Figure 6.12). The received DM-RS allows data to be decoded without knowing the code word used since the receiver receives the effect of MIMO and BF precoding: [H] [X] [W].

The mobile UE transmits a measurement report to the base station from the CSI-RSs.

The gNB base station can transmit CSI-RSs without beamforming. In this case, the matrix X is the identity matrix, also called the single-CSI method.

The gNB base station can transmit CSI-RSs with beamforming, also called the multiple-CSI method.

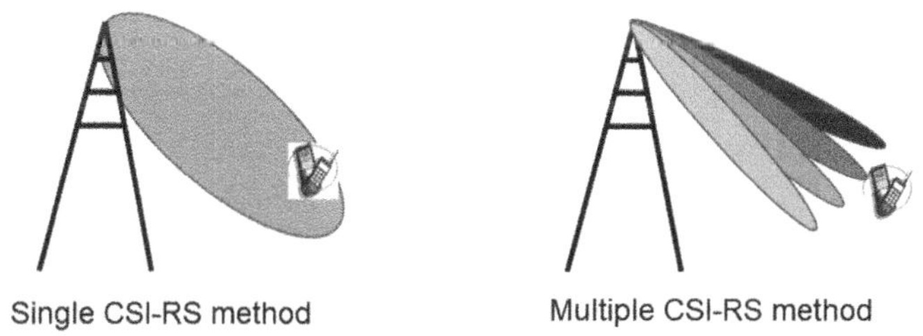

Figure 6.13. *Single CSI-RS and multiple CSI-RS methods*

The massive-MIMO mechanism introduces two methods for the feedback of CSI to the gNB entity: the single CSI-RS method and the multiple CSI-RS method.

The single CSI-RS method is applied to the antenna partitioning model. CSI is collected per antenna element and per polarization of a group from a CSI-RS physical signal, not precoded by the matrix X.

The single CSI-RS method defines a number of antenna ports and a number of non-precoded CSI-RS which can vary from 1 to 32. Single CSI-RS allows horizontal or vertical beamforming and 8x8 MIMO.

The multiple CSI-RS method is applied to the complete connection model, and it consists of reporting the CSI for each beam formed from the CSI-RS.

The multiple CSI-RS method limits the number of antenna ports per beam to 8.

The returned CSI relates to the best CSI-RS received and contains a CRI (CSI Resource Index) corresponding to this CSI-RS.

In the case of the multiple CSI-RS method, the following mechanisms have been introduced to improve efficiency of CSI-RS resources, i.e. to allow more mobiles to share a pool of CSI-RS resources:

– aperiodic CSI-RS, in which a mobile is configured to measure the CSI-RS physical signal in a given sub-frame;

– multi-tap CSI-RS, in which a mobile is configured to measure the periodic CSI-RS physical signal for a specified period of time.

Two-step resource configurations are introduced to allow flexible selection of resources while maintaining control over CSI reporting overload. For these configurations, RRC messages and Medium Access Control (MAC) elements are combined.

The enhancements also include the optimization of CSI reports, obtained from measurements of the CSI-RS physical signal.

The first mechanism aims to configure the size of the return information by defining two types of reports:

– CSI TYPE 1: normal (PMI) feedback in the case of SU-MIMO giving the direction of the most important beam.

– CSI TYPE 2: enhanced (explicit or codebook-based) in the case of MU-MIMO by providing more feedback information from the terminal to the base station.

The second mechanism aims to improve the flexibility of CSI-RS resources, either by reducing the density of the resource elements (see Chapter 4) or by transmitting a measurement of all the beams in order to transmit only the highest power beam (only for the multiple-CSI method).

6.4.2. *Codebook configuration*

The 5G codebook is an evolution of the 4G codebook specified in Release R.13 to implement FD-MIMO.

The 3GPP specification defines a W precoding which consists of two sub-matrices W_1 and W_2 with $W = W_1W_2$. The matrix X is therefore integrated into W.

The matrix W_1 represents the broadband properties of the RF channel and the matrix W_2 a sub-band matrix of co-phasing between the beams.

In the case of beamforming, the weighting vectors are calculated from the angle of incidence of the signal and according to the number of antenna elements:

$$w_k = \frac{1}{\sqrt{K}} \exp\left(-j\frac{2\pi}{\lambda}(k-1)d_v \cos \theta_{etilt}\right)$$

The radiating elements k = 1 to 0_1 are uniformly spaced (ULA: Uniform Linear Antenna) by a distance d_v.

In the case of the codebook, the zenith angle can only take precise values (the codebook is limited in size).

We will assume that the rotation value can only take values $i/0_1$ – i integer values between 0 and 0_1-1. The beam bandwidth and the number of lobes depend on the antenna configuration.

Let (N_g,N_1,N_2) be the structure of the antenna. The antenna elements being composed of a cross-polar strand, the number of CSI-RS signals is $2N_g.N_1.N_2$.

In the case of the single panel, the value of N_g is equal to 1. The values (N_1,N_2) are transmitted to the mobile in an RRC message via the data structure *CodebookConfig -> typeI-SinglePanel-> n1-n2*.

In the case of the multi-panel, the value of N_g is equal to 1. The values (N_1,N_2) are transmitted to the mobile in an RRC message via the data structure *CodebookConfig -> typeI-MultiPanel-> ng-n1-n2*.

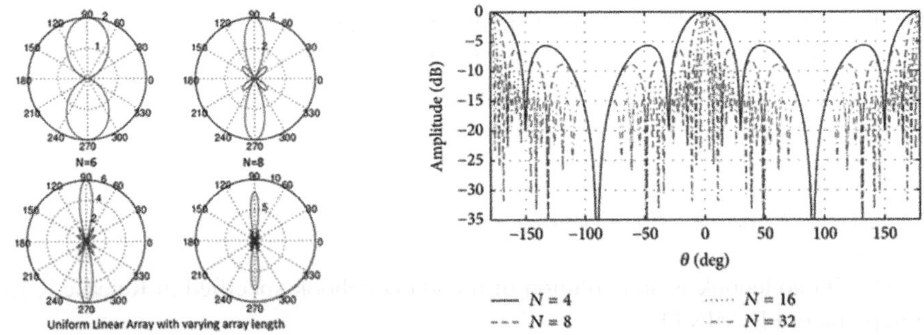

Figure 6.14. *The relationship between the number of lobes, the beamwidth and the number of radiating elements (Su and Chang (2015))*

The matrix W_1 is calculated from the product of a rotation matrix $R_N(q)$ of factor q corresponding to the zenith angle θ_{etilt} and the discrete Fourier transform TFD of a beam. The TFD is defined by the diagonal matrix D.

$$R_N(q) = diag(e^{j\frac{2\pi 0q}{N}}, e^{j\frac{2\pi q}{N}}, ..., e^{j\frac{2\pi(N-1)q}{N}}) \text{ and } [Dn]_{n,m} = \frac{1}{\sqrt{N}} e^{j\frac{2\pi mn}{N}}.$$

$$[R_N(q)Dn]_{n,m} = \frac{1}{\sqrt{N}} e^{j\frac{2\pi m(n+q)}{N}}$$

Beam rotation is equivalent to oversampling the code word (see Discussion R1-1612661 and references) by a factor of $0=0_1.0_2$.

For a 2D antenna, we denote by 0_1 the oversampling of N_1 (horizontal direction) and 0_2 the oversampling of N_2 (vertical direction).

The code word is defined by: $W_1 = \begin{bmatrix} X_1 \otimes X_2 & 0 \\ 0 & X_1 \otimes X_2 \end{bmatrix}$

$$X_1 = \begin{bmatrix} 1 & e^{j\frac{2\pi l}{N_1 O_1}} & e^{j\frac{2\pi(N_1-1)l}{N_1 O_1}} \end{bmatrix}^T \text{ and } X_2 = \begin{bmatrix} 1 & e^{j\frac{2\pi l}{N_{21} O_{21}}} & e^{j\frac{2\pi(N_2-1)m}{N_2 O_2}} \end{bmatrix}^T$$

The coding is applied differently depending on the single- or multi-panel antenna architecture.

The coding configurations are defined in Table 6.4 for single-panel mode and Table 6.5 for multi-panel mode.

Number of CSI-RS ports	N_1,N_2	$0_1,0_2$
4	(2,1)	(4,-)
8	(2,2)	(4,4)
	(4,1)	(4,-)
12	(3,2)	(4,4)
	(6,1)	(4,-)
16	(4,2)	(4,4)
	(8,1)	(4,-)
24	(6,2), (4,3)	(4,4)
	(12,1)	(4,-)
32	(8,2), (4,4)	(4,4)
	(16,1)	(4,-)

Table 6.4. *Codebook for type 1/type 2 single-panel*

Number of CSI-RS ports	N_g,N_1,N_2	$0_1,0_2$
8	(2,2,1)	(4,-)
16	(2,2,2)	(4,4)
	(2,4,1), (4,2,1)	(4,-)
32	(2,4,2), (4,2,2)	(4,4)
	(2,8,1), (4,4,1)	(4,-)

Table 6.5. *Codebook for type 1 multi-panel*

The mobile owns a codebook, which can be defined with the antenna architecture of the base station.

The gNb transmits to the mobile the antenna model via an RRC message in the *CodebookConfig* data structure.

The type 1 codebook corresponds to the case of a single user (SU-MIMO) up to eight spatial layers. Due to the architecture of the antennas, type 1 offers two subtypes: type 1 single-panel CSI and type 1 multi-panel CSI.

The type 2 codebook is dedicated to MU-MIMO transmissions with at most two spatial layers per user.

6.5. PUSCH transmission

Uplink transmission uses either the OFDM signal (Orthogonal Frequency Division Multiplexing) or the DFT-S-OFDM signal (Discrete Fourier Transform Spread OFDM).

Only the OFDM signal allows the MIMO mechanism to operate up to four spatial layers or the beamforming mechanism by digital precoding. MIMO is possible only for mobiles that are able to check the consistency of the signals on the two or four antenna ports.

The MIMO mechanism or the beamforming mechanism for the uplink requires the capacity of the mobile to control the relative phase of the signals transmitted on the different antennas (coherence between antennas). The phase difference between a signal transmitted on a pair of antennas or on the four antennas must be constant (for a wideband signal).

The antenna capacities of the mobile are defined by the following characteristics:

– full coherence: all antenna ports can transmit broadband signals coherently since the mobile is able to control the relative phase of the signals between each antenna;

– partial coherence: some pairs of antenna ports can transmit broadband signals coherently;

– non-coherence: no antenna can transmit broadband signals coherently with another antenna.

Digital precoding involves selecting a precoding code for the uplink.

This code is defined either from the measurements of the uplink channel carried out at the base station, or from the measurements of the downlink channel carried out by the mobile, assuming the reciprocity of the transmission channel.

The SRS allows the base station to estimate the uplink channel before configuring the PUSCH channel transmission mode.

The mobile must be configured to transmit multiple SRSs (SRS multi-port) in order to use the MIMO mechanism.

The CSI-RS allows the mobile to estimate the downlink channel and, by reciprocity, this allows the uplink channel to be considered the same.

Two PUSCH multi-antenna configuration modes are defined as follows:

– transmission based on the codebook (codebook-based) to implement the MIMO or beamforming mechanism;

– transmission not based on the codebook (non-codebook-based) to implement beamforming.

Codebook-based transmission is the mode chosen by the base station when uplink SRS reference signal measurements are required.

For the non-codebook-based transmission mode, the base station sets the number of RI (Rank Indicator) spatial layers (two or four) and indicates to the mobile the index of the PMI code table. The information is carried by the DCI 0 descending control information message and contained in the UL Grant resource allocation request. The mobile will apply the code imposed by the gNB to both the PUSCH channel and the DM-RS reference signal.

For the non-codebook-based transmission, the base station assumes transmission channel reciprocity. In this mode:

1) the mobile selects the code that seems most appropriate for the beamforming mechanism from the CSI-RSs transmitted by the base station;

2) the mobile transmits precoded SRSs to assist the base station in estimating the upstream channel per beam;

3) the code selection done by the mobile aims to select the best code from a mobile point of view. However, this code may cause interference with other users. The base station can modify the choice of code from the mobile measurement report of the SRS-multi-port signals. The code change is then transmitted in the DCI 0 descending control information message and in the UL Grant resource allocation request.

6.6. Beamforming management

6.6.1. *Burst SSB: beam sweeping*

The SSB block requires an allocation of 240 sub-carriers in the frequency domain and four symbols in the time domain.

In the FR1 band, the spacing between sub-carriers of the SSB block is 15 kHz or 30 kHz.

In the FR2 band, the sub-carrier spacing of the SSB block is 120 kHz or 240 kHz.

Numerology	SSB bandwidth	Block duration
0	3.6 MHz	286 μs
1	7.2 MHz	143 μs
3	28.8 MHz	36 μs
4	57.6 MHz	18 μs

Table 6.6. *SSB characteristics*

The SSB is periodically broadcasts by the base station. The periodicity varies between 5 ms and 160 ms. However, the mobile assumes that the periodicity of the SSB is 20 ms.

An SSB burst is made up of several SSBs over a half sub-frame (5 ms):

– 1–4 SSB for a carrier frequency lower than 3 GHz;

– 1–8 SSB for a carrier frequency between 3 GHz and 6 GHz;

– 1–64 SSB in the FR2 band.

Beam sweeping of the synchronization signal is realized by SSB burst broadcasting.

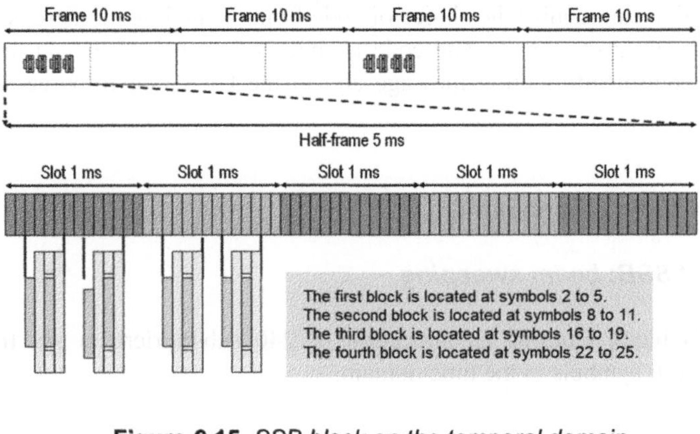

Figure 6.15. *SSB block on the temporal domain*

The periodicity of the SSB block and the number of SSB blocks per burst are defined in the data structure *ServingCellConfigCommonSIB*.

The *ServingCellConfigCommonSIB* is contained in the SIB1 information message for 5G-SA or is transmitted to the mobile during the RRC Connection Reconfiguration request in the case of 5G-NSA.

In the time domain, each SSB block is identified by an SSB index and transmitted in a given direction for beam sweeping.

The mobile measures the power of the signal received at each SSB block and identifies the SSB with the highest power by its index.

6.6.2. *Cell selection and cell re-selection procedures*

The cell selection procedure is initiated by the mobile UE as follows:

– when switching on the terminal;

– after a loss of radio access coverage;

– periodically when the mobile is under the coverage of radio access from an equivalent operator.

The procedure begins by scanning all the radio bands for the synchronization signal. If no synchronization signal is found, the mobile searches for the synchronization signal from a previous RAT (Radio Access Technology).

For the case of 5G, the mobile searches for the SSB, and if it does not detect any signal, the mobile searches for the PSS/SSS 4G synchronization signals.

In the case of 5G-NSA, the mobile searches for the PSS/SSS 4G synchronization signals, and if it does not detect any signal, it searches for the 3G then 2G synchronization signal.

Due to the large bandwidth of the 5G-NR radio band, the scanning time becomes too long if the mobile scans the entire 5G band by frequency hopping of 15 kHz.

To allow fast synchronization, the mobile searches for the frequency number of the GSCN (Global Synchronization Channel Number) by frequency hopping of:

– 1.2 MHz for a carrier frequency lower than 3 GHz;

– 1.44 MHz for a carrier frequency between 3 GHz and 24.5 GHz (Release 16);

– 17.28 MHz in the FR2 band.

Thus, the possibilities for locating the SSB are indicated by parameter N of Table 6.7.

Carrier frequency	Block position SSB (SS_{ref})	GSCN	Frequency raster
f < 3 GHz	N*1200 kHz +M * 50 kHz M={1,3,5} N=1:2499	3N+(M-3)/2	1.2 MHz
3 GHz < f < 24.5 GHz	3000 MHz +N*1.44 MHz N=0:14756	7499+N	1.44 MHz
f > 24.5 GHz	24250.08 MHz +N*17.28 MHz N=0:4383	22256+N	17.28 MHz

Table 6.7. *Frequency search configuration*

Once synchronized (i.e. centered on the GSCN frequency), the mobile is able to determine the frequency position of subcarrier 0 of the SSB.

However, the SSB block must be positioned in a grid of channels spaced 15 kHz apart. The frequency location of an SSB may not be aligned to a PRB. The base station sends information in the MIB message (see Chapter 9) allowing the mobile to know the frequency offset between subcarrier 0 of the SSB block and subcarrier 0 of the PRB block.

This frequency offset is determined by the K_{ssb} parameter.

For the FR1 band, the frequency difference is equal to 15 kHz * K_{ssb}, the value of 15 kHz being the measurement reference.

In the case where the spacing between subcarriers is 15 kHz, the K_{ssb} value is between 0 and 11.

In the case where the spacing between subcarriers is 30 kHz, the K_{ssb} value is between 0 and 23.

For the FR2 band, the frequency deviation is equal to 120 kHz * K_{ssb}.

An additional offset also transmitted in the MIB information message allows the mobile to know the start of the active BWP band.

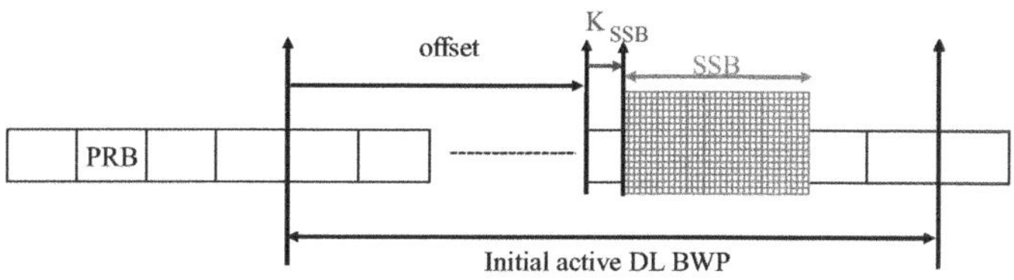

Figure 6.16. *SSB position relative to PRB*

In standby mode, the cell re-selection procedure allows the mobile to select a more suitable cell and to camp on it.

When the base station broadcasts several SSB (beam-sweeping), the mobile measures each SSB. An SSB is identified by an index. The mobile selects the beam on which it can camp and whose reception power is the highest.

6.6.3. *Beam management*

When the mobile wants to establish a radio connection, it begins by sending a random access request, as described in Figure 3.7:

– the mobile UE selects the best SSB beam;

– the mobile UE transmits the random access request in message 1 carrying the preamble;

– the mobile UE optionally repeats the same by increasing the power to the selected beam or to different beams (UE beam sweeping);

– the base station answers the mobile on the beam selected by the mobile (message 2);

– the mobile UE acknowledges the beam selected by message 3;

– the base station confirms the configuration of the radio link.

The random access procedure allows the mobile to notify the radio node of the best radio beam. This beam is dedicated for future exchanges of control messages and data.

Beam management is suitable for CPE (Customer Premise Equipment) equipment such as a 5G/Wi-Fi router (FWA: Fixed Wireless Access).

Mobility requires constant adjustment of the direction of the beams.

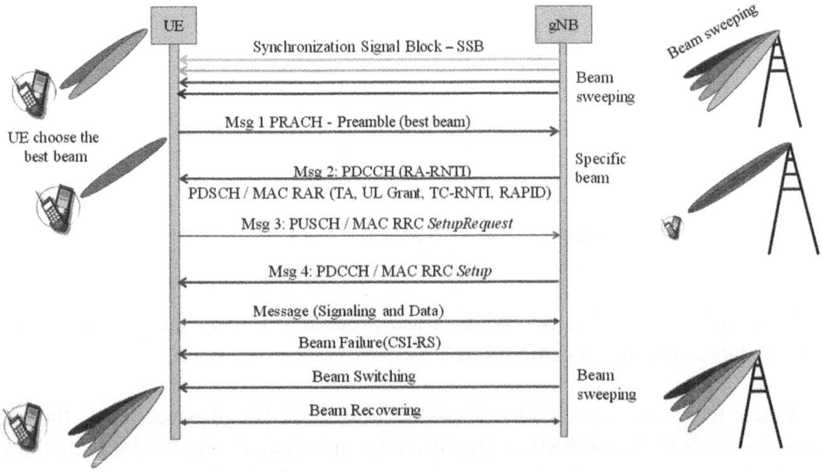

Figure 6.17. *Beam selection procedure*

The beam switching procedure only applies when the mobile is in the intra-cell connected state. This procedure can be transparent to the mobile. The base station requests the mobile to periodically generate CSI measurement reports from the measurements of the CSI-RSs. The base station can therefore switch from one beam to another without informing the mobile.

Beam failure detection is performed by the mobile when measuring the CSI-RSs and/or the SSB if the latter is present in the active band partitioning (BWP).

The MAC layer of the UE mobile has been configured by an RRC message in order to save the beam loss recovery parameters:

– the duration of detection of the loss of the beam;

– the power threshold Q below which the mobile considers the signal to be insufficient;

– the maximum number of measurements for which the measured signal is below the power threshold;

– the power ramp to reach the radio node;

– the maximum number of random access attempts after the beam is lost;

– the preambles for the random access procedure after the beam loss.

In the event of a beam loss, the physical layer transmits an interruption signal to the MAC layer if the L1-RSRP power measurement is below a threshold Q (see Chapter 9).

The MAC layer of the mobile initiates the random access procedure without contention to enable beam failure recovery.

The PRACH procedure is performed in the initial BWP uplink. The mobile listens to the response in the initial BWP band.

The beam loss detection and recovery procedure takes place in the following four stages:

1) the mobile detects a radio beam failure (the reception power measured on the L1-RSRP physical layer is below a threshold);

2) the mobile searches for another beam candidate with the strongest reception power;

3) the mobile starts a random access procedure without contention with the variable *BFR-SSB-Resource.ra-PreambleIndex* as a preamble;

4) the radio node transmits a RACH response with a DCI message.

6.7. References

All standards can be downloaded from the ETSI website: https://www.etsi.org/standards.

3GPP TS 38.211
NR - Physical Channels and Modulation;
Version 16.1.0 Release 16

3GPP TS 38.214
NR - Physical layer procedures for data;
Version 16.2.0 Release 16

3GPP TS 38.215
NR - Physical layer measurements;
Version 16.2.0 Release 16

3GPP TS 38.321
NR - Medium Access Control (MAC) protocol specification;
Version 12.2.0 Release 15

3GPP R1-1612661
TSG-RAN Advanced CSI Codebook Structure
WG1 Meeting 87, 14th November – 18th November 2016

Su, X. and Chang, K.I. (2015). Polarized uniform linear array system: Beam radiation pattern, beamforming diversity order, and channel capacity. *International Journal of Antennas and Propagation* [Online]. Available at: https://doi.org/10.1155/2015/371236.

7

5G-NR Radio Interface – Bandwidth Part

7.1. Bandwidth part

A BandWidth Part (BWP) corresponds to an allocation of resources in the frequency domain.

A BWP part is mobile specific.

Several mobiles can be configured on the same BWP.

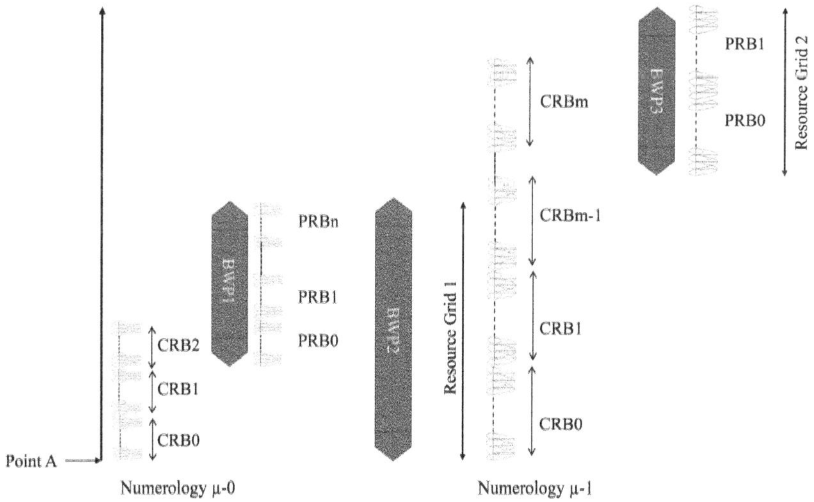

Figure 7.1. *Bandwidth part*

A BWP in a resource grid can overlap.

The bandwidth part is defined by:

– a frequency band corresponding to a number of PRBs (Physical Resource Block);

– a numerology characterizing the bandwidth of the PRB block and the cyclic prefix (normal or extended);

– a starting position calculated by the number of common resource block CRB (Common Resource Block) which separates the BWP band from the reference point A of the 5G-NR carrier.

The initial BWP corresponds to a 5G-NR sub-band on which the mobile UE (User Equipment):

– searches for the SSB (Synchronization Signal Block) (see Chapter 6);

– listens to paging notifications sent to the common search space and to specific control information sent to the dedicated search space;

– issues the random access request for the establishment of the radio connection.

The size of the BWP depends on the number of PRBs and the numerology.

In the standby state, the bandwidth of the BWP is reduced to minimize the power consumption of the UE mobile.

In the connected state:

– the mobile is configured with a BWP active in both directions of transmission (UL: UpLink and DL: DownLink);

– the size of the BWPs can be equal to the maximum of the bandwidth of the 5G carrier;

– multiple BWPs can be aggregated; however, there can only be one active BWP per 5G carrier (one active BWP on the main PCell cell, and one active BWP on each secondary SCell cell).

The BWP makes it possible to adapt to the maximum bandwidth of the terminal, for example IoT (Internet of Things) terminals.

In the standby state, the mobile camps on a cell which broadcasts the SSB block and the SIB1 (System Information Block Type 1) information allowing the mobile to find the configuration of the initial BWP partition.

The configuration of the initial BWP is contained in the Minimum System Information (MSI).

The MSI is transmitted via the MIB (Master Information Block) and SIB1 in the case of 5G-SA or in the RRC *ConnectionReconfiguration* message in the case of 5G-NSA.

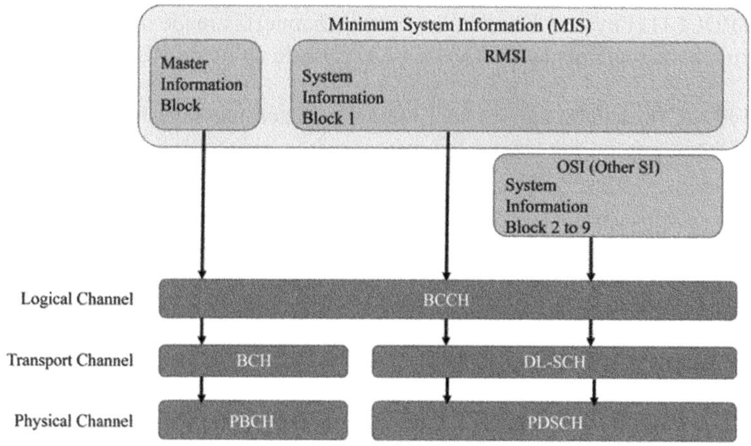

Figure 7.2. *Remaining minimum information system. For a color version of this figure, see www.iste.co.uk/launay/5g.zip*

7.2. CORESET

The concept of CORESET (Control Resource SET) was introduced on the 5G-NR interface to limit the control frequency domain to the capacities of devices.

Unlike the 4G radio interface for which the physical PDCCH (Physical Downlink Control Channel) control channel occupies the entire bandwidth of the carrier, on the 5G-NR interface, the mobile is specifically configured to listen to a subspace of tempo-frequency resources corresponding to the control search space.

DCI (Downlink Control Information) information is transmitted in the CORESET space.

One, two or at most three CORESETs are configured in each BWP.

CORESET is a set of configurable Resource Elements (RE). The configuration is stored in the *ControlResourceSET* element.

The Resource Elements (RE) in a CORESET are grouped together by RBs (Resource Blocks), called the Resource Element Group (REG).

The PDCCH physical channel and the DM-RS (Demodulation-Reference Signal) reference signals associated with the physical channel are transmitted in a CORESET.

The PDCCH (Physical Downlink Control Channel) is made up of an aggregation of control channel elements (CCEs). A CCE is made up of six REGs.

The PDCCH channel carries DCI messages. Each message is defined by a DCI format. Depending on the size of the message, it is necessary to aggregate 1, 2, 4, 8 or 16 CCEs.

The CORESET is formed by CCEs grouped into RB (REG) associated with the DM-RS reference signals. REGs can be interlaced to provide frequency diversity or can be contiguous for beam localization (beamforming).

The CORESET is defined:

– in the frequency domain by a number of resource blocks;

– in the time domain by a number of symbols;

– by the number of REGs present;

– by the level of aggregation L corresponding to the number of CCEs allocated to the PDCCH.

The configuration of the CORESET (*ControlResourceSET IE*) is transmitted to the mobile UE via RRC messages.

7.2.1. *Configuration of CORESET#0*

Each CORESET is located in a carrier part named BWP. When the mobile switches on, it searches for the initial BWP band and listens to DCI messages carried by the PDCCH channel in the CORESET.

CORESET#0 is associated with the initial BWP band configured from broadcast signals.

The MIB broadcast message contains:

– the SFN (System Frame Number);

– the numerology (*subCarrierSpacingCommon*) of the initial BWP;

– the shift in the frequency domain between the position of the SSB block and the start of the resource grid (*ssb-SubcarrierOffset*) in the number of sub-carriers;

– the position of the first reference signal DM-RS (*dmrs-TypeA-Position*);

– the bandwidth of the control space of the PDCCH channel (*pdcch-ConfigSIB1*);

– the state of the cell (prohibited or not: CellBarred);

– authorization of intra-band cell re-selection (intraFreqReselection).

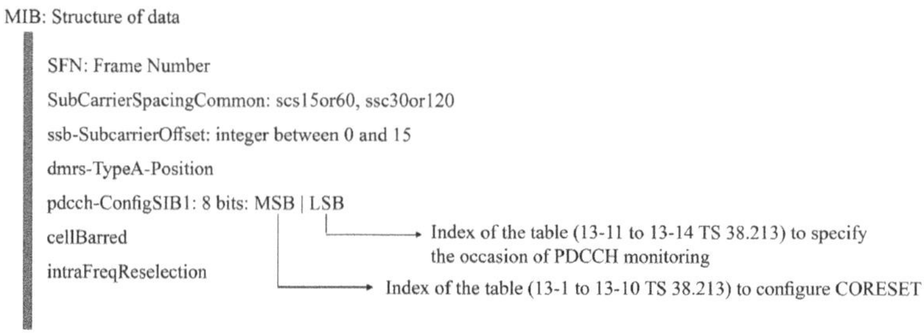

Figure 7.3. *Data structure of MIB. MSB: Most Significant Bit, LSB: Least Significant Bit*

The initial BWP is a configuration common to all UE mobiles which camp in the cell. It is possible to limit access to the cell.

The value *subCarrierSpacingCommon* takes as its value:

– 15 kHz in the FR1 band or 60 kHz in the FR2 band (Scs15or60);

– 30 kHz in the FR1 band or 120 kHz in the FR2 band (Scs30or120).

The *ssb-SubcarrierOffset* parameter designates the offset between the first subcarrier of the SSB block and the common block resource. This shift comes from the search for cells by frequency hopping (see Chapter 4 – frequency raster).

The dmrs-TypeA-Position parameter indicates the position in the time domain of the first DM-RS reference signal associated with the PDSCH channel (pos2 or pos3, see Chapter 4).

The pdcch-ConfigSIB1 parameter will allow the mobile to know the position of the PDCCH physical control channel and the area of the common search space. The common search space allows the gNB (gNode B) entity to transmit common information to all the mobiles in the cell. The search space is located inside the CORESET control area.

The *pdcch-ConfigSIB1* parameter is an 8-bit encoded value:

– the first 4 bits (most significant bits) determine the index of the initial CORESET#0 control area. The index allows us to know the size of CORESET#0 in the frequency domain and the number of symbols in the time domain. CORESET#0 carries the Type0-PDCCH common search space;

– the last 4 bits (least significant bits) determine the index of search zone 0, allowing the mobile to know the times when it must scan the PDCCH channel (PDCCH monitoring occasion).

When selecting the cell, the mobile searches for the presence of the Type0-PDCCH common search space in CORESET#0.

The index of the initial CORESET#0 control area corresponds to the row of a table allowing the following information to be stored by the mobile:

– the number $N_{RB}^{CORESET}$ of RB;

– the number of symbols $N_{SYMB}^{CORESET}$;

– the offset in number of RBs between the start of the grid and the position in the frequency domain of the search zone.

Index MSB	Number of RBs	Number of symbols	Offset (RB)
0	24	2	0
1	24	2	1
2	24	2	2
3	24	2	3
4	24	2	4
5	24	3	0
6	24	3	1
7	24	3	2
8	24	3	3
9	24	3	4
10	48	1	12
11	48	1	14
12	48	1	16
13	48	2	12
14	48	2	14
15	48	2	16

Table 7.1. *Extract from Table 13.4 TS 38.213*

The possible configurations of CORESET#0 for the initial BWP defined from the decoding of the pdcch-ConfigSIB1 parameter are described in Tables 13-1 to 13-10 of the TS 38.213 specification.

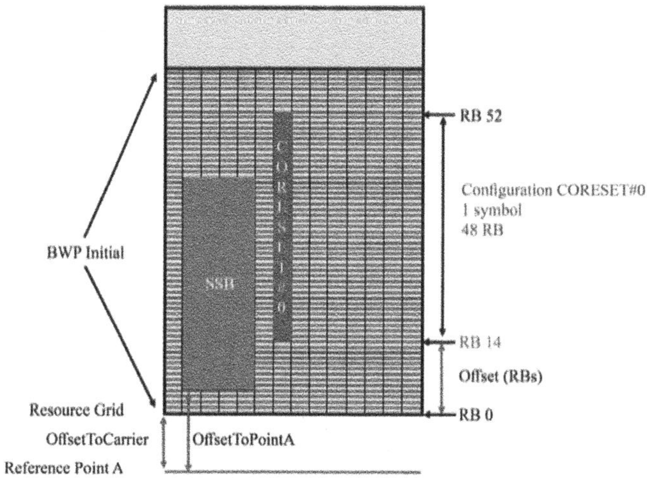

Figure 7.4. *CORESET#0 configuration with index 14. For a color version of this figure, see www.iste.co.uk/launay/5g.zip*

The position of CORESET#0 in the time domain is obtained by the values of TS 38.213 Table 13-11 whose row is calculated from the four least significant bits.

CORESET#0 carries the sequencing information of the SIB1 system information. Once the MIB information has been decoded, the mobile will be able to recover the information transmitted by the broadcast message SIB1.

The SIB1 broadcast message contains the Remaining Minimum System Information (RMSI):

– the cell selection criteria;

– the identity of the PLMN (Public Land Mobile Network);

– the code of the TAC (Tracking Area Code);

– the identity of the cell;

– the unified access control of the UAC (Unified Access Control) cell;

– the RAN (Radio Access Network) notification information;

– the resource allocation information for OSIs (Other System Information);

– the common information of the service cell (*ServingCellConfigCommon*).

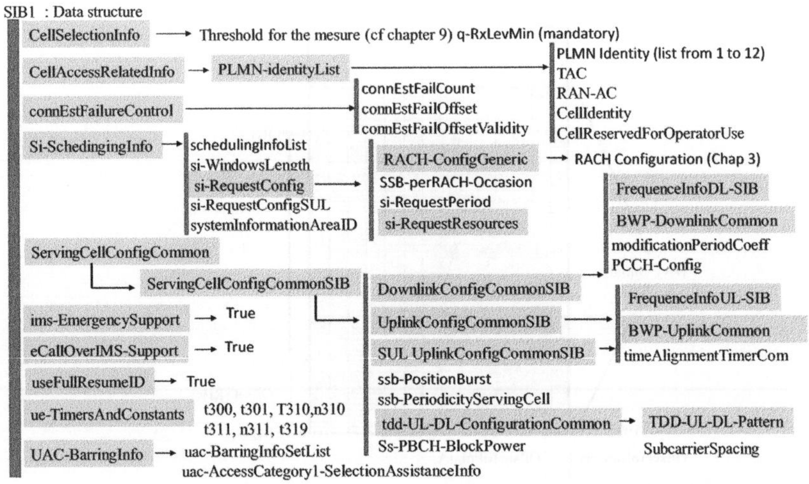

Figure 7.5. *SIB1 data structure. For a color version of this figure, see www.iste.co.uk/launay/5g.zip*

Unified Access Control (UAC) is a set of mechanisms that determine whether to allow or deny a mobile to camp on the cell depending on the service requested and the load of the cell.

The information in the *ServingCell_ConfigCommon* data block is used to configure the initial BWP of the cell.

For the uplink, the configuration is defined in the *BWP-UplinkCommon* data block.

For the downlink, the configuration is defined in the *BWP-Downlink* data block.

The *BWP-Downlink* data block refers to two other data structures:

– the *BWP-DownlinkCommon* data block used to configure the common parameters of the cell applicable to all terminals;

– the *BWP-DownlinkDedicated* data block used to configure the parameters specific to a mobile.

The RMSI allows the mobile to be located on the initial BWP and be configured to receive CORESET#0 by broadcasting MIB and SIB1.

7.2.2. CORESET configuration

The configuration of a new CORESET and the control space are contained in the *PDCCH-Config* IE (Information Element). The new CORESET is specific to the mobile.

The network configures up to three CORESETs (CORESET#0 included) per BWP and per cell.

For 5G-SA, the configuration of a new CORESET is transmitted in the RRC *Reconfiguration* message during the connection establishment procedure.

Information	**The data structure where parameters are contained**
DL BWP	*spCellConfigCommon.frequencyInfoDL* *spCellConfigCommon.InitialDownlinkBWP*
UL BWP	*spCellConfigCommon.frequencyInfoUL* *spCellConfigCommon.InitialUplinkBWP*
PRACH configuration	*spCellConfigCommon.uplinkConfigCommon.initialUplinkBWP.rach-ConfigCommon*
PDCCH configuration	*spCellConfigCommon.initialDownlinkBWP.pdcch-ConfigCommon*
Dedicated PDCCH configuration	*spCellConfigDedicated.initialDownlinkBWP.pdcch-Config*
PDSCH configuration	*spCellConfigCommon.initialDownlinkBWP.pdsch-ConfigCommon*
Dedicated PDSCH configuration	*spCellConfigDedicated.initialDownlinkBWP.pdsch-Config*
PUSCH configuration	*spCellConfigCommon.uplinkConfigCommon.initialUplinkBWP.PUSCH-ConfigCommon*
Dedicated PUSCH configuration	*spCellConfigDedicated.initialDownlinkBWP.pdcch-Config*
TDD UL/DL configuration	*spCellConfigCommn.tdd-UL-DL-ConfigurationCommon* *spCellConfigDedicated.tdd-UL-DL-ConfigurationDedicated*

Table 7.2. *Radio configurations*

For 5G-NSA, the configuration of CORESET#0 is transmitted by the master node (eNB (evolved Node B) entity) during the connection reconfiguration procedure for dual connectivity in the RRC *ConnectionReconfiguration* message.

The information carried by the RRC *Reconfiguration* or RRC *ConnectionReconfiguration* message is described in Table 7.2 (*SpCellConfig* data structure).

In the time domain, a CORESET can occupy up to three OFDM symbols, contiguous or not. If the sequencing is done by slot, it is preferable to position the CORESET at the start of the slot (see Chapter 4 – DM-RS Mapping Type A). For low latency communications, the CORESET can be positioned on any symbol of the slot (see Chapter 4 – DM-RS Mapping Type B).

7.3. BWP switching procedure

The initial BWP is transmitted to the mobile by the broadcast information MIB and SIB1.

The initial BWP configuration is cell specific for categories of mobile, and the dedicated control space resources are configured by CORESET#0 through the presence of the Type0-PDCCH search space.

For operations on the PCell primary cell, the mobile is configured for an initial uplink band (*initial-UL-BWP*) and an initial downlink *band-DL-BWP* to implement the random access procedure (see Chapter 3).

The gNB entity transmits the BWP configuration to the mobile:

– during the establishment of the radio link, when the mobile goes from the standby state to the connected state;

– or during the reconfiguration of the radio link.

The configuration is transmitted in the RRC *Reconfiguration* message, and the mobile is positioned on a single BWP band, called active BWP.

In the connected state, at most four BWPs are configured at the mobile level in order to be able to switch from one part to another quickly.

In the Release 15 and Release 16 specifications, the mobile is only allowed to position itself on a single BWP, the mobile is configured at a given time with a single numerology.

BWP switching is triggered by a control message transmitted from the gNB entity to the mobile or at the end of a timer:

– a DCI format 1_1 or format 0_1 control message;

– the expiration of the BWP-InactivityTimer (between 2 ms and 2 s).

The RRC message is used for the configuration of the radio link.

The DCI message allows dynamic adaptation of the resource allocation by telling the mobile to switch to another BWP Uplink or Downlink part outside of the original BWP.

The mobile switches to the initial BWP when the BWP-InactivityTimer expires.

In the event of loss of the radio link or of the beam (see Chapter 6), the mobile switches to the initial BWP to be able to start the random access procedure.

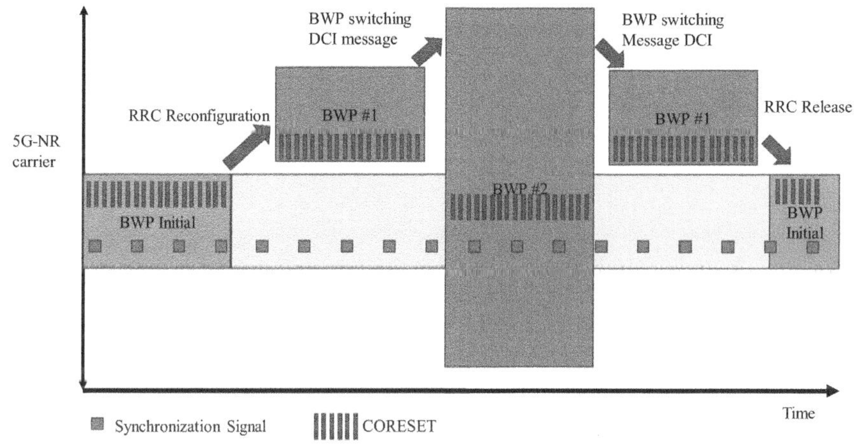

Figure 7.6. *BWP switching. For a color version of this figure, see www.iste.co.uk/launay/5g.zip*

BWP switching is controlled by the physical PDCCH channel or by the MAC (Medium Access Control) layer during the random access procedure.

In case of carrier aggregation, a BWP is activated by default when activating the SCell (secondary cell) in the RRC *Reconfiguration* message without requiring a DCI control message.

In case of dual DC-NR connectivity, a BWP is enabled by default when activating the secondary SpCell (Special Cell) in the RRC *Reconfiguration* message without requiring a DCI control message.

7.4. References

All standards can be downloaded from the ETSI website: https://www.etsi.org/standards.

3GPP TS 38.211
NR - Physical Channels and Modulation;
Version 16.1.0 Release 16

3GPP TS 38.213
NR - Physical layer procedures for control;
Version 16.2.0 Release 16

3GPP TS 38.306
NR - User Equipment (UE) radio access capabilities;
Version 16.1.0 Release 16

3GPP TS 38.331
NR - Radio Resource Control (RRC);
Version 15.2.1 Release 15

8

5G-NR Radio Interface – Data Link Layer

The data link layer is made up of four sublayers:

– SDAP (Service Data Adaptation Protocol);

– PDCP (Packet Data Convergence Protocol);

– RLC (Radio Link Control);

– MAC (Medium Access Control).

A level n sublayer provides services to the n + 1 layer.

Each level n layer receives data from the upper layer which it transfers to the lower layer by inserting control information in a header H.

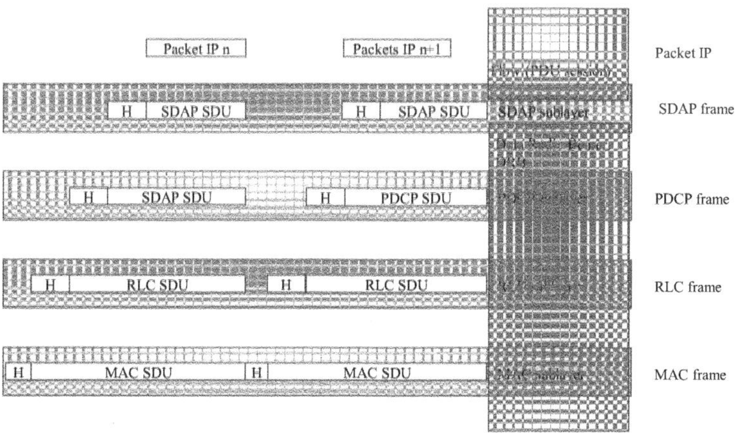

Figure 8.1. *Data link layer protocol. For a color version of this figure, see www.iste.co.uk/launay/5g.zip*

The data received by the upper sublayer is referred to as Service Data Unit (SDU).

The frame transmitted to the lower sublayer comprises the SDU data to which control information is added.

8.1. SDAP protocol

Connectivity to a data network through 5G-NR radio access and the 5G Core network (5GC) is provided by a PDU (Packet Data Unit) session.

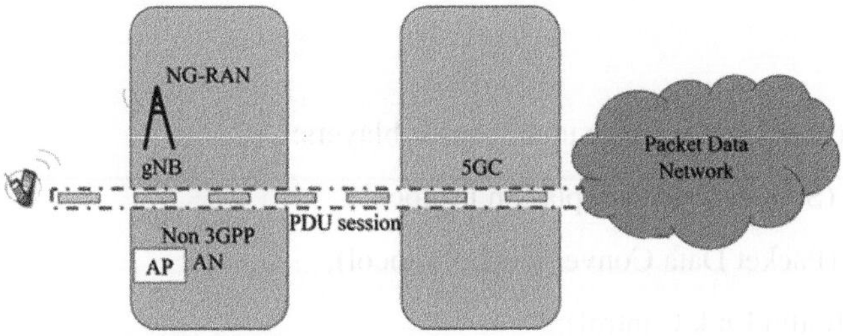

Figure 8.2. *PDU session. For a color version of this figure, see www.iste.co.uk/launay/5g.zip*

The PDU session is defined by:

– the format of the data to be transported (Ethernet, IP, unstructured);

– the SMF (Session Management Function) involved in the establishment of the sessions;

– the type of service in an S-NSSAI (Single – Network Slice Selection Assistance Information).

The PDU session consists of:

– a tunnel in the core network between the UPF (User Plane Function) and the gNB (gNode B) entity via the N3 interface;

– one or more radio supports (bearer).

The QoS (Quality of Service) model is based on flow management. User traffic is classified into Service Data Flows (SDF). An SDF service data flow refers to a group of flows associated with a service.

For the 4G mobile network, QoS management is controlled at the bearer level. The EPS bearer is a tunnel that groups together different SDFs that have the same QoS. A TFT (Traffic Flow Templates) filter makes it possible to group together the service flows towards the corresponding bearer. The TFT filter is configured by the operator according to five parameters: source and destination IP address, source and destination port, and the identity of the protocol.

For the 5G mobile network, the QoS flow is characterized by a QoS profile provided by the SMF function to the NG-RAN radio access.

The QoS profile contains a set of rules for determining the processing of the flow on the radio interface:

– class of service identifier (5QI);

– the allocation and retention priority ARP (Allocation and Retention Priority).

The 5QI class of service identifier corresponds to the index of a table on which the flow characteristics are predefined:

– priority;

– the flow transfer mechanism: flow at guaranteed rate (GBR) or not guaranteed (non-GBR);

– the rate of packet loss;

– the RTT time relating to the transfer of the flow in an end-to-end PDU session.

Unlike 4G, for which end-to-end connectivity (EPS bearer) was characterized by a unique QoS, the 5G PDU session carries flows, and each flow is characterized by its own QoS. Thus, different QoS flows are transported in the same PDU session.

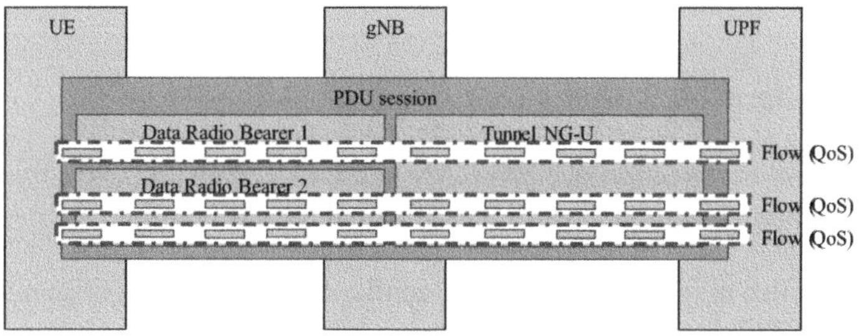

Figure 8.3. *PDU session and QoS flow. For a color version of this figure, see www.iste.co.uk/launay/5g.zip*

At the NAS (Non Access Stratum) level, the QoS flow in a PDU session is identified by a QFI (QoS Flow Identifier).

The QoS profile is managed at the NAS level (ARP + QCI (QoS Class Identifier)):

– the maximum packet loss characteristics are configured only for multimedia streams at guaranteed GBR rate (voice, videoconferencing);

– the Reflective QoS Attribute (RQA) only concerns non-GBR unsecured rate flows. The RQA attribute is optional.

If the SMF transmits a request for a QoS parameter control notification relating to certain GBR flows to the base station, then the base station informs the SMF function when the actual throughput is lower than the minimum expected value or when the minimum throughput is reached again.

At the AS (Access Stratum) level, the matching rules are based on the QFI parameter. QFI rules allow us to associate the stream with the corresponding DRB radio medium at both the uplink direction (mobile's rules) and downlink direction (base station's rules).

Different DRB media can be configured for each stream with different packet routing specifications.

Several streams belonging to the same PDU session can be multiplexed in the same DRB bearer.

Only one SDAP entity is configured by the RRC layer per PDU session except for the dual connectivity mechanism or two SDAP entities are configured.

Several SDAP entities can be simultaneously configured for a mobile.

8.1.1. *Operations*

In the downlink direction, the NG-RAN radio node maps the QoS streams to the DRB radio bearers based on the QFI identifier and the QoS profile.

The QoS rules at the mobile level are configured during the configuration of the radio link by RRC signaling messages.

The mobile applies QoS or reflective QoS for the correspondence between the flow and the bearer.

Reflective QoS allows the UE to create its QoS rules from the downlink traffic without additional signaling.

The SDAP protocol provides the following operations:

– the mapping between the data flows exchanged with the core network contained in a tunnel, and the Quality of Service to be applied on the radio interface on the uplink and downlink;

– marking the identity of the QoS flow in the upstream and downstream packets;

– the implementation of reflective QoS on the correspondence of flows in the uplink direction.

8.1.1.1. *UE side operations*

For each PDU session, a default radio bearer is configured.

The SDAP entity is configured by the RRC control layer. The configuration allows us to create mapping rules between the identifier QFI of the flow and the corresponding radio medium (bearer). The mapping rules are saved by the mobile until the SDAP entity is released.

The mapping rules between the QoS flow and the DRB is controlled by the radio node:

– from the RRC signaling which configures the mapping rules between the flow QoS and the radio bearer;

– from the mobile which applies the same rules for uplink traffic as for downlink traffic (reflective QoS).

Any stream transmitted to the SDAP layer is:

– in the absence of a matching rule, sent without adding an SDAP header to the default bearer;

– in the presence of matching rules, the flow is routed to the PDCP sublayer of the bearer with or without addition of an SDAP header.

The RRC configuration is used to define:

– the addition or not of an SDAP header by bearer;

– the implementation of QoS or reflective QoS.

To achieve this mapping, it is necessary:

– that the base station transmits the QFI indicator to the mobile during a previous exchange;

– that the base station informs the mobile to apply reflective QoS.

For dual connectivity, the QoS flow of the same PDU session can be transmitted in different DRBs. One SDAP entity is configured for the master node and another for the secondary node.

8.1.1.2. *gNB side operations*

Packets coming from the UPF are encapsulated in a tunnel and transmitted to the gNB via the N3 interface. The QFI identifier is carried by a header extension of the GTP-U (GPRS Tunnel Protocol – User Plane) protocol. The gNB base station associates each QFI flow of a PDU session with the corresponding DRB radio bearer.

In the downlink direction, to apply the reflective QoS at the mobile level, the RQI (Reflective QoS Indication) is set to 1 (see Figure 8.3). The RQI identifier is transmitted from the UPF to the gNB entity in the header extension of the GTP-U protocol. The SDAP header of the frame transmitted by the base station contains the QFI identifier.

The SDAP frame is transmitted, for each bearer, to the PDCP sublayer corresponding to the bearer.

In the uplink direction, the base station applies the routing rules to transmit the IP packet towards the UPF.

8.1.2. *The protocol structure*

The structure of the protocol concerns the transmission and reception function. Both are implemented at the gNB and at the UE.

The SDAP entity transmits the flows to the lower layers. Depending on the NAS configuration, the entity SDAP adds or not a header (see Figure 8.4).

The SDAP entity receives the frames from the PDCP layer.

Figure 8.4 on the left describes the SDAP header added by the base station for the transmission of traffic towards the UE and the reception of traffic from the UE.

Figure 8.4 on the right describes the SDAP header added by the UE for transmission towards the gNB and reception from the gNB.

The SDAP entity is not compelled to add an SDAP header.

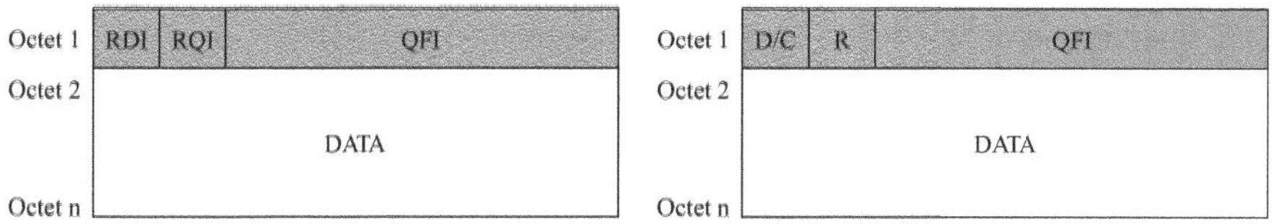

Figure 8.4. *SDAP frame structure*

RDI (Reflective QoS flow to DRB mapping Indication): when this bit is coded to 1, the mobile updates the matching rule.

RQI (Reflective QoS Indication): when this bit is coded to 1, the mobile informs the NAS layer to update the matching rules.

D/C (Data/Control): this bit indicates whether the frame contains SDAP frames (bit at 1) or control messages (bit at 0).

8.2. PDCP

PDCP (Packet Data Convergence Protocol) is used for RRC (Radio Resource Control) signaling messages, relative to dedicated control data, and for traffic Internet Protocol (IP) packets.

The PDCP provides the following functions:

– compression of traffic data headers using the ROHC (Robust Header Compression) mechanism;

– the security of RRC traffic and signaling data (integrity and confidentiality);

– sequence numbering and sequential delivery of RRC messages and IP packets;

– the recovery of PDCP frames lost during the handover;

– reestablishment of PDCP frames in the event of a loss of the RLC frame in acknowledged mode;

– duplication of PDCP PDUs;

– detection of duplication;

– deletion of PDCP frames (discard);

– routing of PDCP frames in the case of dual connectivity (split-bearer).

Several PDCP instances can be activated simultaneously:

– two instances for SRB1 (Signaling Radio Bearer) and SRB2:

- the SRB1 is used for the transmission of an RRC message that can support a NAS (Non-Access Stratum) message,

- SRB2 is used for transmission of a NAS message only;

– one instance for each DRB (Data Radio Bearer) relating to traffic data.

8.2.1. *Procedures*

8.2.1.1. *Header compression*

The header compression is based on the ROHC mechanism, for which multiple algorithms have been defined by the Request For Comment (RFC) specifications of the Internet Engineering Task Force (IETF) standards body (see Table 8.1).

Profile identifier	Compressed header	References
0x0000	no compression	RFC 4995
0x0001	RTP/UDP/IP	RFC 3095, RFC 4815
0x0002	UDP/IP	RFC 3095, RFC 4815
0x0003	ESP/IP	RFC 3095, RFC 4815
0x0004	IP	RFC 3843, RFC 4815
0x0006	TCP/IP	RFC 4996
0x0101	RTP/UDP/IP	RFC 5225
0x0102	UDP/IP	RFC 5225
0x0103	ESP/IP	RFC 5225
0x0104	IP	RFC 5225

Table 8.1. *Compression protocols*

The header compression is based on the observation that in a session, a number of fields are invariant, such as IP addresses or port numbers.

The header compression is particularly effective when the size of the payload is relatively low, which is the case for voices (Figure 8.5).

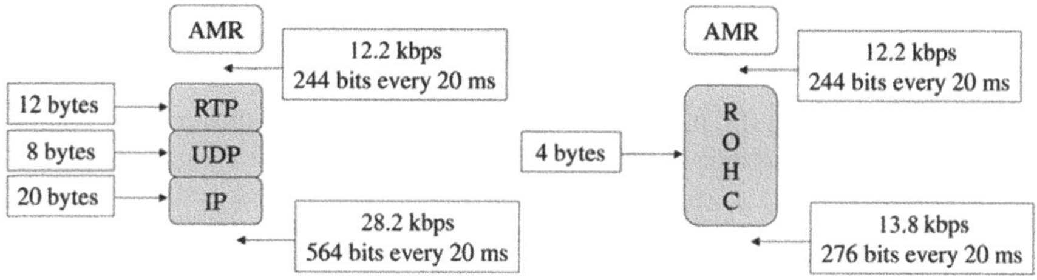

Figure 8.5. *Header compression. AMR: Adaptive Multi-Rate*

The ROHC decompressor uses the PDCP Feedback control message to inform the ROHC compressor that the decompression is successful or that the synchronization between compression and decompression has been lost.

8.2.1.2. *Loss of frames during handover*

The PDCP allows the recovery of frames that had been lost during handover, but only if the Radio Link Control (RLC) protocol uses the acknowledgement mode.

For the downlink direction, during the handover procedure, the source gNB entity sends the target gNB entity an Xn-AP *SN STATUS NUMBER* message in which it provides, on the one hand, the PDCP frames transmitted to the mobile and unacknowledged with their Sequence Number (SN) and, on the other hand, the Sequence Numbers (SN) of non-received PDCP frames from the mobile for the uplink direction.

When the mobile connects to the target gNB entity, the recovery of lost frames is performed by the following exchanges:

– the mobile sends the target gNB entity the PDCP Status Report message which mentions the missing PDCP frames for the downlink direction;

– the target gNB entity sends the PDCP Status Report message to the mobile which mentions the missing PDCP frames for the uplink direction.

8.2.1.3. *Packet removal (PDCP discard)*

During the handover or when the radio link is not good enough, the maximum data rate reached at the radio interface is lower than the rate desired by the application.

Buffering PDCP packets is used to resolve the jitter issue. Buffering data can also temporarily compensate the reduction in data rate.

Deleting packets at the transmitter allows the transmission buffer to be emptied, especially in the case of real-time applications.

A frame drop timer is activated for each PDCP SDU frame. When the timer expires, if the PDCP frame is not transmitted, the latter frame is deleted.

8.2.1.4. *Split bearer*

The split bearer is a function to allow the split of data going to a node (MN or SN) from the core side into two paths, the 1st path through the air interface (MN or SN) towards the UE and the 2nd path towards the X2/Xn interface with the other node (respectively SN or MN) which will transfer this data through the air interface towards the UE.

Traffic splitting is performed by the PDCP entity and is transmitted to the RLC entities of the master cell and the secondary cell.

The MR-DC (Multi-RAT Dual Connectivity) is triggered by the MN (master node) which transmits the order to add a secondary SN node via the X2C/X2n control interface.

Several configurations are possible:

– dual EN-DC connectivity (E-UTRA – NR-DC): the master node is the eNB entity, and the secondary node is the gNB entity (5G-NSA option 3);

– dual NE-DC connectivity (NR-DC – E-UTRA): the master node is the gNB entity, and the secondary node is the eNB entity;

– NGEN-DC dual connectivity (E-UTRA – NR-DC): the master node is the ng-eNB entity, and the secondary node is the gNB entity.

Each radio node has its own RRC entity (E-UTRA RRC for the eNB base station and NR RRC for the gNB base station).

The SRB signaling radio bearer carries RRC messages over the radio interface.

The initial configuration of the secondary SN node (RRC message) is transmitted via the X2-AP or Xn-AP protocol from the secondary SN node to the master node (MN) through the X2-C/Xn-C interface.

For 5G-NSA, option 3, the SRB1 signaling bearer supports the establishment of the initial PDCP E-UTRA connection.

The subsequent signaling messages:

– SRB1 and SRB2 are transmitted by the PDCP E-UTRA entity or the PDCP NR entity from the master node to the mobile. The SN RRC messages from the gNB entity are transmitted to the master eNB node, and encapsulated in an MN RRC message;

– SRB3 is transmitted by the PDCP NR entity from the secondary node to the mobile, allowing the RRC messages from the gNB entity to be transported directly to the mobile.

The SRB split is supported for SRB1 and SRB2 media. The RRC messages are encrypted and fully protected by the NR-PDCP entity of the master node (MN).

The support separation takes place:

– option 3: at the level of the master node eNB;

– option 3x: at the level of the secondary node gNB.

Each node is responsible for allocating resources for its cell.

8.2.1.5. *PDU duplication*

PDU duplication is supported for the user plane and for the control plane.

Duplication is configured at the RRC layer. Duplication involves transmitting the PDCP frame multiple times through different RLC sublayers. The primary and secondary RLC layers are activated for each DRB radio bearer.

In case of dual connection, the following configurations are defined in the Release 16 specification:

– EN-DC: the duplication can be applied to the aggregation of carriers at the level of the master node and at the level of the secondary node but only under the control of the E-UTRAN PDCP entity and provided that there is no DC duplication.

– NGEN-DC or NE-DC: the duplication can only be configured at the level of the gNB entity, either at the level of the secondary support SCG (Secondary Cell Group) bearer for the NGEN-DC option or at the level of the MCG (Master Cell Group) bearer master support.

– NR-DC: duplication takes place on both radio nodes.

On reception, the duplicated packets are deleted (discard).

8.2.2. *Operations*

8.2.2.1. *Operations relating to the SRB bearer*

Figure 8.6 describes the operations relating to the SRB bearer.

At transmission, the RRC messages are subject to a reorder procedure which helps the detection of duplicate messages on the receiver side.

The Message Authentication Code for Integrity (MAC-I) allows integrity control of the PDCP frame.

The calculation of the MAC-I is done with the following information:

– the PDCP frame that contains the PDCP header and RRC message;

– the COUNT parameter that aggregates the Hyper Frame Number (HFN) and the PDCP frame sequence number;

– the DIR parameter which indicates whether the data are transmitted in the downlink or uplink direction;

– the BEARER parameter which is the identity of the bearer (5 bits);

– the K_{RRCint} key.

The RRC message and the MAC-I authentication code are then encrypted using the K_{RRCenc} key, then encapsulated by the PDCP header to form the PDCP frame transmitted to the RLC layer.

On receiving the PDCP frames from the RLC layer, the following operations occur:

– the PDCP header is removed;

– the RRC message and the MAC-I authentication code are decrypted;

– the RRC message and the PDCP header are integrity controlled;

– the RRC messages are delivered in sequence.

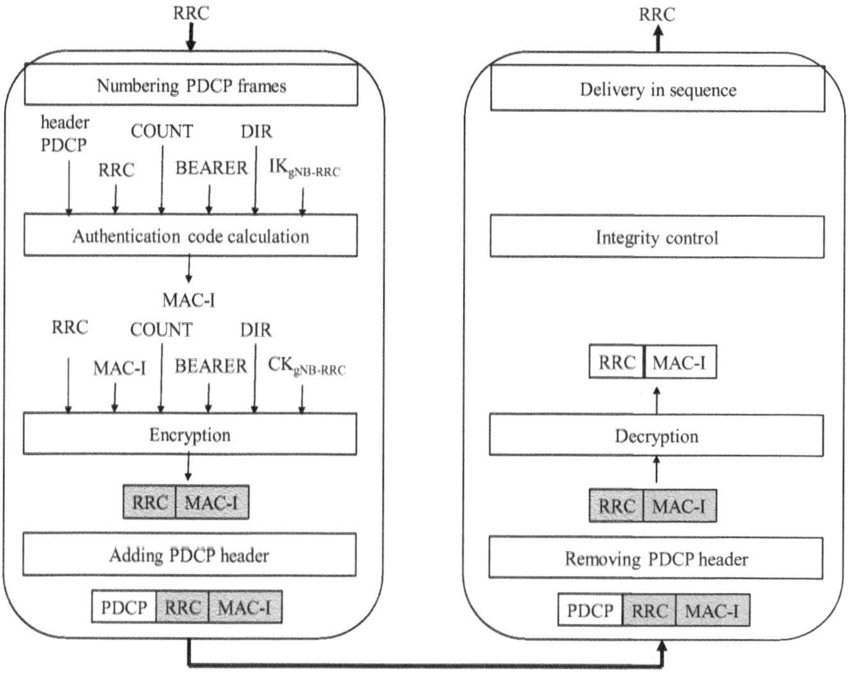

Figure 8.6. *PDCP operations relating to the SRB bearer*[1]

8.2.2.2. *Operations relating to the DRB bearer*

Figure 8.7 describes the operations relating to the DRB bearer.

At transmission, the IP packets are numbered which allows the receiver to detect duplicate or lost frames and to order.

The sequence number is used during handover and allows retransmission requests for lost packets, but only if the acknowledgment mode is used for the RLC layer.

The IP packet may be subject to the ROHC header compression to reduce the relative weight of headers.

1 In all figures, the shaded blocks are encrypted.

The Message Authentication Code for Integrity (MAC-I) allows integrity control of the PDCP frame. Integrity for data is optional.

The calculation of the MAC-I message authentication code is done from the following information:

– the PDCP frame that contains the PDCP header and RRC message;

– the COUNT parameter that aggregates the Hyper Frame Number (HFN) and the PDCP frame sequence number;

– the DIR parameter that indicates whether the data are transmitted in the downlink or uplink direction;

– the BEARER parameter which is the identity of the bearer (5 bits);

– the K_{UPint} key.

The IP packet is encrypted using the K_{UPenc}.

If the messages are also integrity protected, then the MAC-I seal is encrypted using the K_{UPint}, before being encapsulated by the PDCP header to form the PDCP frame forwarded to the RLC layer.

On receiving the PDCP frames from the RLC layer, the following operations occur:

– the PDCP header is removed;

– the seal is verified;

– the IP packet is decrypted;

– the IP packet is optionally decompressed;

– the duplicate PDCP frames are removed and the IP packets are delivered in sequence.

The PDCP feedback control message checks the header compression procedure.

The PDCP Status Report control message allows recovery of the frames lost during the handover.

The control messages of the PDCP, Control PDCP PDU, are not protected in integrity.

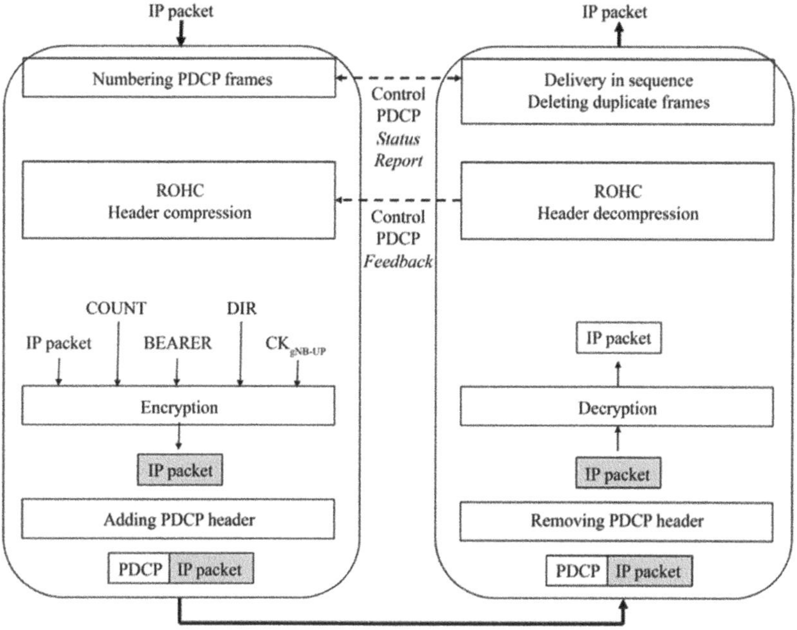

Figure 8.7. *PDCP operations relating to the DRB bearer*

8.2.3. *Protocol structure*

The PDCP defines headers to encapsulate the RRC signaling data, the data traffic and the control messages associated with the traffic data.

The structure of NR-PDCP frames is described in Figure 8.8.

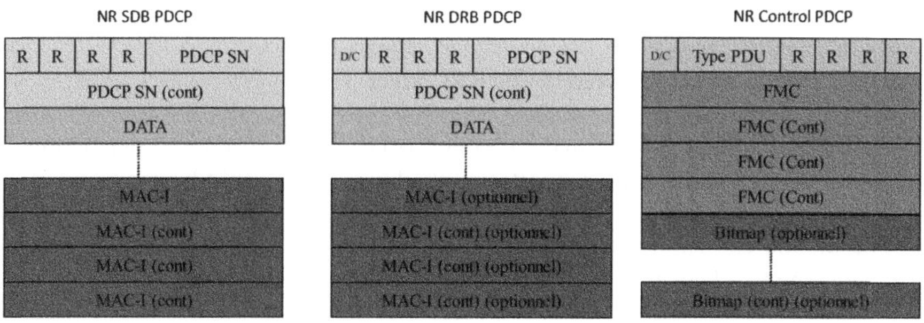

Figure 8.8. *NR-PDCP frame structure. For a color version of this figure, see www.iste.co.uk/launay/5g.zip*

PDCP SN: this field is coded on 12 bits for RRC signaling data and on 12 or 18 bits for traffic data. It indicates the PDCP frame sequence number. The sequence numbering allows recovery of the PDCP frames lost during the handover.

MAC-I: this field is coded on 4 bytes. It contains the authentication code to control the integrity of the PDCP frame containing RRC signaling data and optionally the IP traffic.

Data/Control (D/C): this bit indicates whether the frame contains traffic data (bit value ONE) or control messages specific to the PDCP (bit value ZERO).

Packet Data Unit (PDU) type: this field is coded on 3 bits. It shows the control message type associated with traffic data:

– the Status Report (value 000) is used in the reestablishment process after a cell change. It indicates a list of PDCP frame numbers not received during the handover;

– the ROHC PDCP feedback (value 001) message is linked to the ROHC header compression mechanism.

First Missing (FMS): this field is coded on 12 bits. It contains the first sequence number of the missing PDCP frames.

Bitmap: this field is a collection of bits indicating whether the PDCP frame was received correctly (bit value ONE) or not (bit value ZERO). The most significant bit of the first byte represents the sequence number following the value of the FMS field.

8.3. RLC protocol

The Radio Link Control (RLC) protocol provides control of the radio link between the mobile and the gNB entity.

The RLC entity is configured by the RRC layer.

Each PDCP entity is associated with one, two or four RLC entities depending on the characteristics of the radio support: uplink, downlink and split bearer.

The mobile can simultaneously activate multiple RLC instances, each instance corresponding to a Packet Data Convergence Protocol (PDCP) instance.

The RLC protocol operates in three operation modes:

– Acknowledged Mode (AM);

– Unacknowledged Mode (UM);

– Transparent Mode (TM), for which no header is added to the data.

The RLC protocol does the following:

– retransmission in the case of error via the Automatic Repeat reQuest (ARQ) mechanism, for the acknowledged mode only;

– segmentation and reassembly of PDCP frames in both the acknowledged and unacknowledged modes;

– possible re-segmentation of PDCP frames, in acknowledgment mode, during a retransmission of the RLC frame;

– re-sequencing of received data, in both the acknowledged and unacknowledged modes;

– detection of duplicate data in both the acknowledged and unacknowledged modes.

Figure 8.9 presents the operating modes of the RLC protocol used by the various flow types.

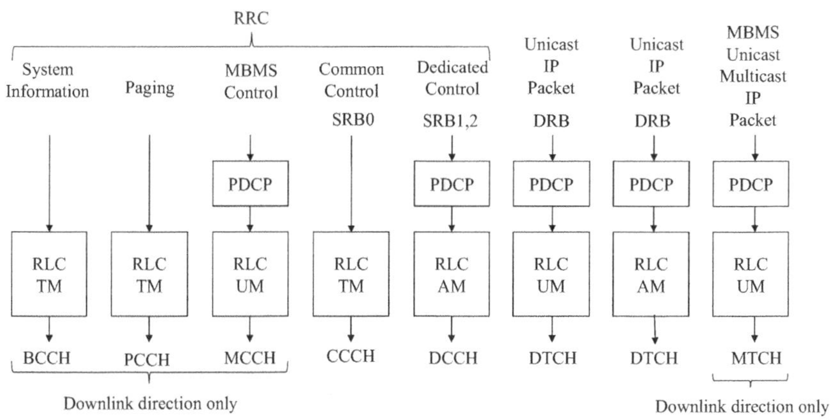

Figure 8.9. *Operating mode of the RLC protocol. MBMS: Multicast Broadcast*

8.3.1. *Operations*

8.3.1.1. *TM mode*

At transmission, the Radio Resource Control (RRC) messages are stored in a memory, before being transmitted to the Medium Access Control (MAC) layer.

At reception, the data received from the MAC layer are transmitted directly to the RRC layer.

Figure 8.10 describes the operations performed for the transparent mode.

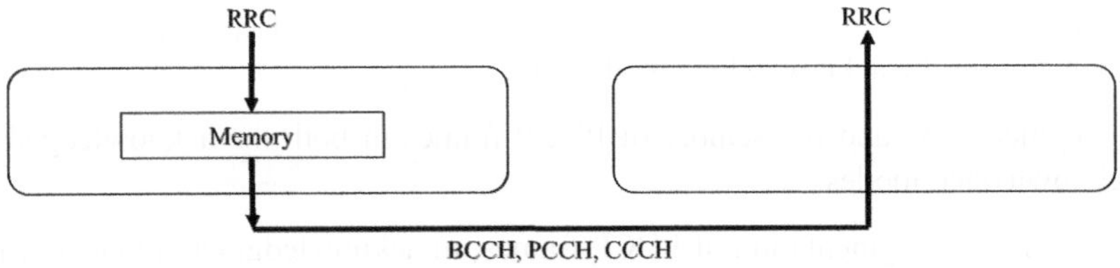

Figure 8.10. *TM mode operations*

8.3.1.2. *UM mode*

At transmission, the PDCP frames are stored in a memory.

When transmitting to the MAC layer, the PDCP frames may be segmented, and are then encapsulated by an RLC header.

At reception, the data received from the MAC layer are stored in a memory.

The received data are ordered, following desequencing which can be introduced by the Hybrid Automatic Repeat reQuest (HARQ) mechanism, and the RLC header is then removed.

The last operation is to reassemble the segments to reconstruct the original PDCP frame.

Figure 8.11 describes the operations for the unacknowledged mode.

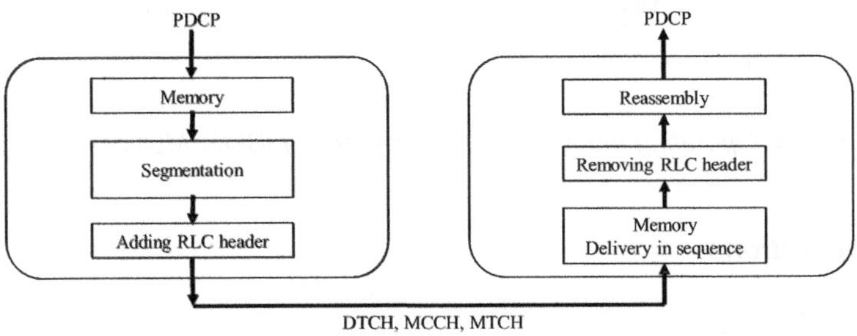

Figure 8.11. *UM mode operations*

8.3.1.3. *AM mode*

At transmission, the PDCP frames are stored in a memory.

When transmitting to the MAC layer, the PDCP frames may be segmented, and are then encapsulated by an RLC header.

The RLC frames transmitted are stored in a retransmission memory.

The data received from the MAC layer are stored in a memory and are placed in sequence, and the RLC header is then removed.

The last operation is to reassemble the segments to reconstruct the original PDCP frame.

A request may be sent from the source to the destination to receive the status report of an RLC frame reception.

An RLC control message is sent from the destination to the source for the positive or negative acknowledgment of RLC frames.

If the RLC frame is not received correctly, the source will perform a retransmission.

If the Modulation and Coding Scheme (MCS) of the physical layer has been modified, it is possible that the size of the RLC frame can be reduced for the retransmission, and in this case, it is necessary to perform a re-segmentation.

Figure 8.12 demonstrates the operations for the acknowledged mode (AM).

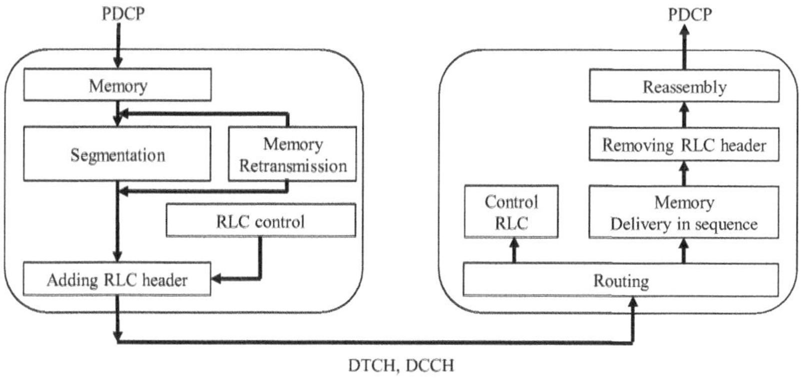

Figure 8.12. *AM mode operation*

8.3.2. *Protocol structure*

The RLC protocol defines a header for the UM mode, a header for the AM mode and a control message.

8.3.2.1. *UM mode*

The header structure depends on the size of the Sequence Number (SN) field and the number of concatenated PDCP frames.

Figure 8.13 demonstrates the RLC header for the following three cases:

– a 6-bit SN field or 12-bit SN field, without offset;

– a 6-bit SN field or 12-bit SN field, with offset;

– no header added when frames are transmitted to MAC without segmentation.

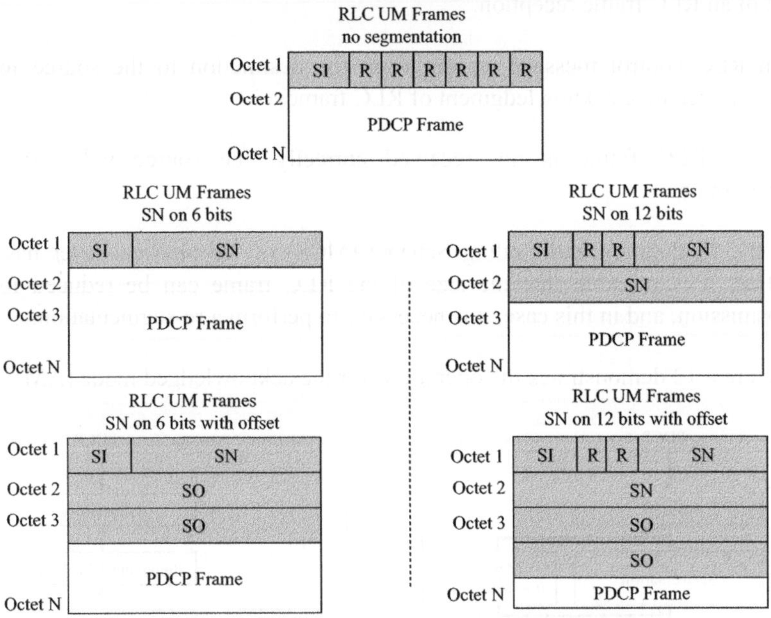

Figure 8.13. *RLC frame structure – UM mode*

SI (Segmentation Information): this field is coded on 2 bits. It indicates whether segmentation is implemented or not:

– SI = 00, no segmentation;

– SI = 01, start of segmentation of the last PDCP frame;

– SI = 10, end of segmentation of the first PDCP frame;

– SI = 11, end of segmentation of the first PDCP frame and start of segmentation of the last PDCP frame.

SO (Segment Offset): this field indicates the position of the RCL SDU segment. The unit of SO is in bytes.

SN: this field is coded on 6 or 12 bits. It contains the RLC frame sequence numbers. The numbering of RLC frames is used to re-sequence the frames received from the MAC sublayer.

R: reserved bit.

8.3.2.2. *AM mode*

The structure of the RLC header consists of a header and PDCP data. The SN field, in terms of size, is 12 bits or 18 bits.

Figure 8.14 demonstrates the RLC header for the two lengths of the SN without segment offset.

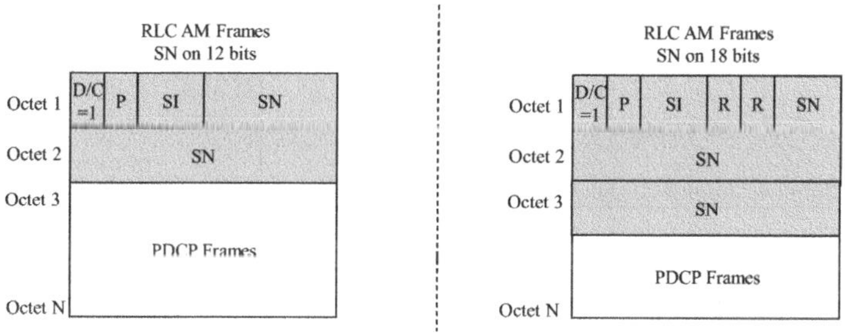

Figure 8.14. *RLC structure of frame – AM mode*

Data/Control (D/C): this bit indicates whether the frame contains PDCP frames (bit value ONE) or control messages (bit value ZERO).

P (Polling): this bit indicates whether the transmission of an acknowledgment report for received frames is required (bit value ONE) or not (bit value ZERO).

SI (Segmentation Information): this indicates whether retransmission is performed with segmentation or without:

– SI = 00, no segmentation;

– SI = 01, start of segmentation of the last PDCP frame;

– SI = 10, end of segmentation of the first PDCP frame;

– SI = 11, end of segmentation of the first PDCP frame and start of segmentation of the last PDCP frame.

In the case of retransmission of an RLC frame for which the original resource is not available, the originally transmitted PDCP frame must be able to be segmented.

Figure 8.15 demonstrates the RLC header in the following two cases with a segment offset, for an SN on 12 bits or 18 bits:

– retransmission of a PDCP frame as a segment;

– retransmission of three PDCP frames, with the first PDCP frame as a segment.

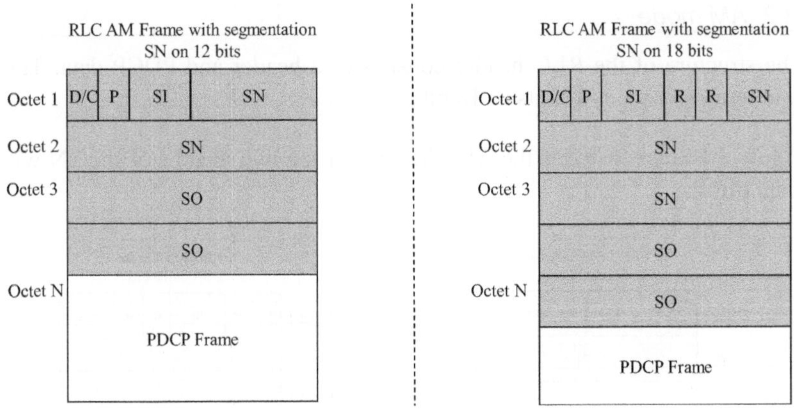

Figure 8.15. *Structure of an RLC frame – AM mode*

SO (Segment Offset): this field indicates the position of the initial PDCP segment. The unit of SO is in bytes.

8.3.2.3. *Control message*

The control message concerns the RLC frame acknowledgement and the indication of lost frames and frame segments.

Figure 8.16 demonstrates the RLC protocol control message either with a sequence numbering of 12 bits (left) or 18 bits (right).

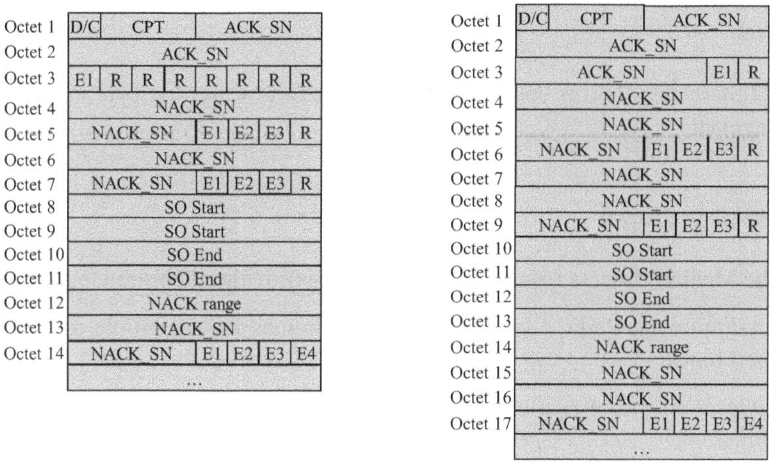

Figure 8.16. *RLC protocol control message*

ACK_SN (Acknowledgement SN): this field is coded on 10 bits. It indicates the number of the next expected frame. All transmitted frames, whose number is less than this value, are validated, with the exception of frames whose number is in the NACK_SN field.

CPT (Control PDU Type): this field is coded on 3 bits. It indicates the control message type. The only defined message concerns the report relating to the acknowledgement of received RLC frames.

E1 (Extension 1): this bit indicates the presence (bit value ONE) or not (bit value ZERO) of NACK_SN, E1 and E2 fields following one another.

E2 (Extension 2): this bit indicates the presence (bit value ONE) or not (bit value ZERO) of SO Start and SO End fields following one another.

E3 (Extension 3): this bit indicates the presence (bit at ONE) or not (bit at ZERO) of a PDCP sequence following the one received.

NACK_range: this field contains the number of consecutive PDCP frames that have been lost.

NACK_SN (Negative Acknowledgement SN): this field is coded on 10 bits. It contains the number of the lost frame.

SO End: this field is coded on 15 bits. It indicates the last byte number of the lost segment.

SO Start: this field is coded on 15 bits. It indicates the first byte number of the lost segment.

8.4. MAC protocol

The Medium Access Control (MAC) protocol provides the following functions:

– multiplexing Radio Link Control (RLC) frames from multiple instances in a transport block;

– allocation of the resources via a scheduling mechanism for both transmission directions;

– management of retransmission in the case of error via the Hybrid Automatic Repeat reQuest (HARQ) mechanism.

8.4.1. *Operations*

8.4.1.1. *gNB side operation*

The operations at the gNB entity level are demonstrated in Figure 8.17.

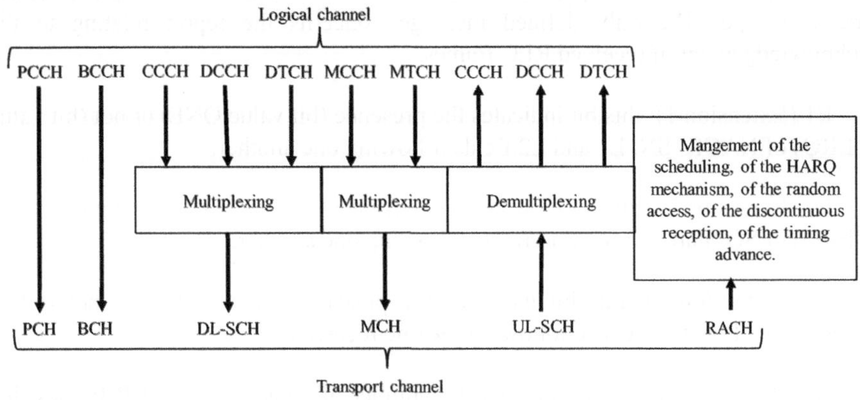

Figure 8.17. *MAC operation gNB side*

The MAC layer transfers the Paging Control Channel (PCCH) to the Paging Channel (PCH).

The MAC layer transfers the Master Information Block (MIB) messages from the Broadcast Control Channel (BCCH) to the Broadcast Channel (BCH).

The MAC layer performs multiplexing, for the downlink direction, of different logical channels to the Downlink Shared Channel (DL-SCH).

The different logical channels multiplexed for the downlink direction are the following:

– the BCH logical channel containing System Information Block (SIB) messages;

– the Common Control Channel (CCCH) containing Radio Resource Control (RRC) common control messages;

– the Dedicated Control Channel (DCCH) containing the RRC dedicated control messages;

– the Dedicated Traffic Channel (DTCH) containing Internet Protocol (IP) packets.

The MAC layer performs demultiplexing, for the uplink direction, of the Uplink Shared Channel (UL-SCH) to restore the various logical channels.

The different logical channels resulting from demultiplexing for the uplink direction are the following:

– the CCCH logical channel containing the RRC common control messages;

– the DCCH logical channel containing the RRC dedicated control messages;

– the DTCH logical channel containing the IP packets.

The MAC layer performs multiplexing, for the downlink direction, of the Multicast Control Channel (MCCH) and Multicast Traffic Channel (MTCH) to the Multicast Channel (MCH).

The MAC layer control entity performs the following functions:

– management of the scheduling of the different logical channels for both directions of transmission;

– management of the HARQ retransmission mechanism;

– management of the mobile random access on reception of the Random Access Channel (RACH);

– management of Discontinuous Reception (DRX);

– management of Timing Advance (TA).

8.4.1.2. *UE side operation*

The operations at the User Equipment (UE) level are described in Figure 8.18.

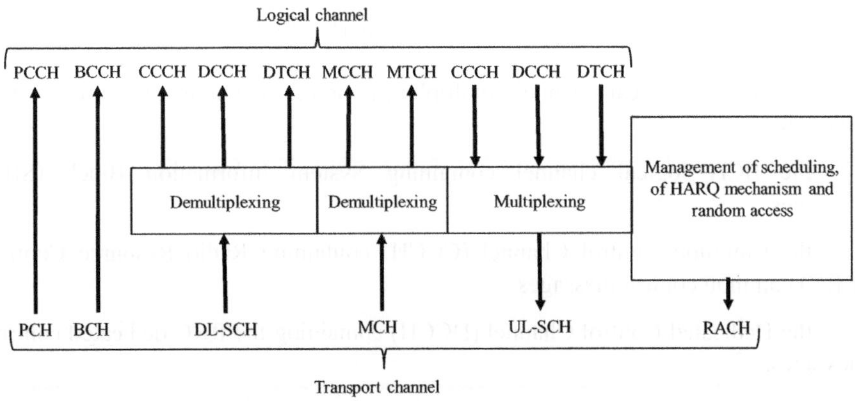

Figure 8.18. *MAC Operation: UE side*

The MAC layer transfers the PCH transport channel to the PCCH logical channel.

The MAC layer transfers the MIB messages of the BCH transport channel to the BCCH logical channel.

The MAC layer performs multiplexing, for the uplink direction, of different logical channels to the UL-SCH transport channel.

The different logical channels multiplexed for the uplink direction are as follows:

– the CCCH logical channel containing the RRC common control messages;

– the DCCH logical channel containing the RRC dedicated control messages;

– the DTCH logical channel containing the IP packets.

The MAC layer performs demultiplexing, for the downlink direction, of the DL-SCH transport channel to restore the different logical channels.

The different logical channels resulting from demultiplexing for the downlink direction are as follows:

– the BCH logical channel containing the SIB messages;

– the CCCH logical channel containing the RRC common control messages;

– the DCCH logical channel containing the RRC dedicated control messages;

– the DTCH logical channel containing the IP packets.

The MAC layer performs demultiplexing, for the downlink direction, of the MCH transport channel to restore the MCCH and MTCH logical channels.

The MAC layer control entity performs the following functions:

– management of the scheduling of different logical channels in the resource allocated by the eNB entity;

– management of the HARQ retransmission mechanism;

– generation of the RACH transport channel.

8.4.2. *Protocol structure*

The MAC protocol defines a header for multiplexing RLC frames from different instances and from MAC protocol control elements.

In 4G, the global MAC protocol header is made up of a collection of unit headers, each unit header being relative to an RLC frame or a control element. RLC data was multiplexed into the MAC payload.

In 5G, the MAC header is placed immediately before RLC data or before MAC control element.

Figure 8.19 describes the general MAC frame structure. Each header unit has a size of 2 or 3 bytes depending on the size of the field length.

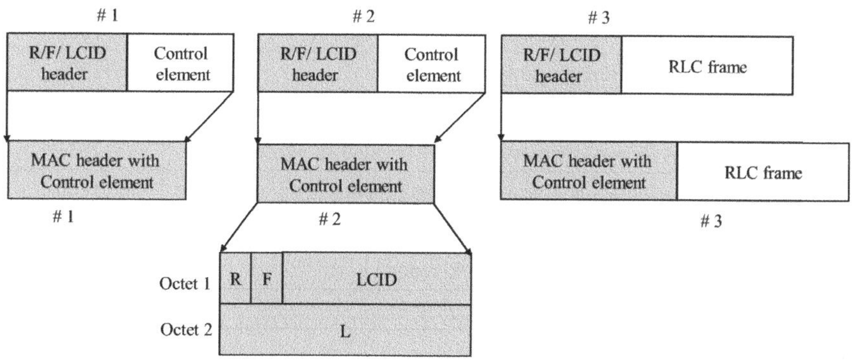

Figure 8.19. *Structure of MAC frame (L in octet)*

F (Format): this bit specifies the Length (L) field format, coded on 7 bits (bit value ONE) or 15 bits (bit value ZERO).

LCID (Logical Channel Identifier): this field is coded on 5 bits. It indicates the logical channel instance identifier or the type of control element (Tables 8.2–8.4).

The identifiers assigned to the DCCH and DTCH logical channels are defined during the connection procedure.

Description	Value
CCCH logical channel identifier	000000
DCCH and DTCH logical channel identifiers	000001–100000
Reserved	100001–101111
ADM Activation/Deactivation	110000–111010
DRX discontinuous reception	111011–111100
Timing Advance (TA)	111101
UE Contention Resolution Identity (CRI)	111110
Filling	11111

Table 8.2. *LCID field values for the DL-SCH transport channel*

Description	Value
CCCH logical channel identifier	000000
Logical channel identifier	000001–100000
Reserved	100001–110110
Configuration of allocation confirmation	110111
PHR reserve power available (extended)	111000
PHR reserve power available	111001
C-RNTI (Cell Radio Network Temporary Identifier)	111010
BSR (Buffer Status Report) memory status (truncated short report)	111011
BSR memory status (truncated long report)	111100
BSR memory status (short report)	111101
BSR memory status (long report)	111110
Filling	111111

Table 8.3. *LCID field values for the UL-SCH transport channel*

Description	Value
MCCH logical channel identifier	000000
MTCH logical channel identifier	00001–11100
Reserved	11101
MSI	11110
Filling	11111

Table 8.4. *LCID field values for the MCH transport channel*

In case of LCID values for the DL-SCH channel, the activation or deactivation concerns:

– ZP CSI-RS (Zero Power CSI-RS) resource elements (LCID = 110000);

– transmission diversity for the PUCCH channel (LCID = 110001);

– SRS (Sounding Reference Signal) resource elements (LCID = 110010);

– the CSI report transmitted by the PUCCH channel (LCID = 110011);

– indication of the TCI (Transmission Configuration Indicator) transmission state for the PDSCH channel (LCID = 110101);

– all CSI-RS/CSI-IM resources (LCID = 110111);

– duplication (LCID = 111000);

– the SCell (secondary cell) (LCID = 111001–111010).

The MAC Random Access Response (RAR) frame is a specific frame used by the gNB entity in response to receiving a random access attempt by the mobile.

Figure 8.20 describes the general structure of an MAC RAR frame with a delay indication.

BI (Backoff Indicator): this field coded on 4 bits provides the value of the delay between two consecutive transmissions of the preamble on the Physical Random Access Channel (PRACH).

Extension (E): this bit indicates whether the next header unit is present (bit value ONE) or not (bit value ZERO).

Random Access Preamble Identifier (RAPID): this field coded on 6 bits contains the identifier of the preamble used by the mobile for random access to the gNB entity.

Type (T): this bit provides an indication of the header unit structure (T = 0 for BI and T = 1 for RAPID).

The RAR message contains the following information:

– Timing Advance (TA) which enables mobile synchronization for the uplink direction;

– Allocation of a resource (UL Grant) for the uplink direction so that the mobile can transmit the following RRC messages:

- *Request* for connection establishment,

- *ReestablishmentRequest* for connection re-establishment,

- *ReconfigurationComplete* when a handover occurs;

– Temporary Cell Radio Network Temporary Identifier (TC-RNTI) allocated to the mobile.

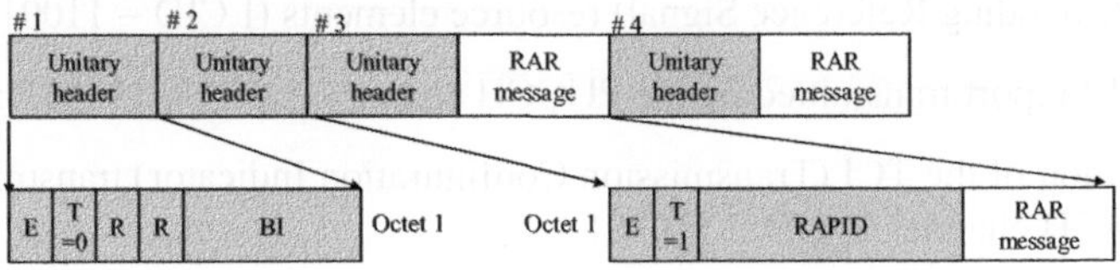

Figure 8.20. *MAC RAR structure of frame*

8.4.3. *Control element*

8.4.3.1. *BSR control element*

The Buffer Status Report (BSR) control element is transmitted by the mobile to provide the memory status in order to obtain the resource for transmission in the Physical Uplink Shared Channel (PUSCH).

The BSR control element can be transmitted periodically, when new and priority data are stored in the memory, or when the size of the MAC frame filling field is greater than that of the control element.

The short and truncated report provides the status of a single memory and contains two fields:

– the Logical Channel Group (LCG) ID field identifies the group of logical channels assigned to that memory;

– the Buffer Size field contains an index corresponding to the size of data available in the memory.

The long report provides the status of all the memories and contains the four Buffer Size fields corresponding to the four logical channel groups.

8.4.3.2. *C-RNTI control element*

The Cell Radio Network Temporary Identifier (C-RNTI) control element contains the identifier assigned to the mobile during random access.

The C-RNTI control element is transmitted by the eNB entity simultaneously with the RRC *ConnectionSetup* message.

8.4.3.3. *DRX control element*

The Discontinuous Reception (DRX) control element enables discontinuous reception to increase the battery life.

If the mobile is idle, in the RRC-IDLE state, discontinuous reception allows it to avoid analyzing all of the Physical Downlink Control Channel (PDCCH) to detect the presence of paging.

If the mobile is connected, in the RRC-CONNECTED state, discontinuous reception allows it to avoid analyzing all the PDCCH physical channels to detect the presence of data in the Physical Downlink Shared Channel (PDSCH) or power commands for the Physical Uplink Control Channel (PUCCH) or for the PUSCH physical channel.

8.4.3.4. *UE CRI control element*

The UE Contention Resolution Identity (CRI) control element is transmitted by the gNB entity simultaneously with the RRC *Setup* message.

The UE CRI control element contains either the Shortened Temporary Mobile Subscriber Identity (S-TMSI) if the mobile is attached to the Attach and Mobility Management Function (AMF), or a random value otherwise.

As in the allocation of the TC-RNTI identifier, several mobiles can consider themselves to be the holder, the UE CRI control element then resolves the contention.

8.4.3.5. *TA control element*

The Timing Advance (TA) control element contains the timing advance value transmitted by the gNB entity.

The TA control element ensures that data transmitted from several mobiles are synchronized at the gNB entity level.

8.4.3.6. *PHR control element*

The Power HeadRoom (PHR) control element contains the indication of the mobile power reserve, the difference between maximum power and power used for the PUSCH physical channel.

The PHR control element is periodically transmitted by the mobile, the periodicity being indicated in the RRC *Setup* or *Reconfiguration* message transmitted by the gNB entity.

The PHR control element is also transmitted when the variation of the attenuation due to propagation is greater than a threshold indicated in the same RRC messages.

Release 10 introduced a control element that is used to indicate the power reserve for each of the aggregated Component Carriers (CCs).

8.4.3.7. *MSI control element*

The MCH Scheduling Information (MSI) control element indicates scheduling information of the Multicast Channel (MCH).

The MSI control element indicates the end of the MTCH physical channel if the MCH transport channel allocated more data than the MTCH physical channel requires.

8.4.3.8. *ADM control element*

The Activation/Deactivation MAC (ADM) control element was introduced in Release 10 and concerns activation and deactivation of the SCell secondary radio channels.

The ADM control element is used when radio channels have been previously established by an RRC *ConnectionSetup* or *ConnectionReconfiguration* message.

The control element allows us to reduce UE power consumption, since rapid deactivation of an SCell radio channel allows the mobile to avoid processing relating to that channel.

8.5. References

All standards can be downloaded from the ETSI website: https://www.etsi.org/standards.

3GPP 23.501
System architecture for the 5G System (5GS);
Version 15.9.0 Release 15

3GPP 37.324
E-UTRA, NR - Multi-connectivity; Overall Description;
Version 15.3.0 Release 15

3GPP 37.340
LTE, 5G, NR - Service Data Adaptation Protocol (SDAP); Protocol specification;
Version 16.0.0 Release 16

3GPP 38.300
NR - Radio Resource Control (RRC); Protocol specification;
Version 16.0.0 Release 16

3GPP 38.321
NR - Medium Access Control (MAC) Protocol specification;
Version 15.3.0 Release 15

3GPP 38.322
NR - Radio Link Control Protocol (RLC) specification;
Version 15.3.0 Release 15

3GPP 38.323
NR - Packet Data Convergence Protocol (PDCP) specification;
Version 16.1.0 Release 16

3GPP 38.331
NR - Radio Resource Control (RRC); Protocol specification;
Version 16.0.0 Release 16

9

5G-NR Radio Interface – Radio Access Procedure

9.1. System information

System information broadcasts radio characteristics of the cell, radio characteristics of neighboring cells and alerts.

System information consists of a Master Information Block (MIB) message and a series of System Information Block (SIB) messages.

The MIB and SIB1 constitute the minimum information for the mobile. SIB1 carries RMSI (Remaining Minimal System Information).

Table 9.1 provides the characteristics of the MIB message support and SIB messages.

SRB	RLC mode	Logical channel	Transport channel	Physical channel
No object	TM	BCCH	BCH	PBCH
Master Information Block				
No object	TM	BCCH	DL-SCH	PDSCH
System Information Block				

Table 9.1. *Support of message relating to system information*

The BCCH logical channel transports MIB information on 24 bits.

The PBCH physical channel adds an additional 8 bits of payload (PBCH Payload).

Generation of the PBCH data is therefore composed of a 24-bit block from the BCH channel, then 8 bits and the 24-bit CRC code (Figure 9.1).

The payload of the PBCH contains:

– 4 least significant bits completing the frame number transmitted in the MIB message;

– 3 bits for (depending on the number of SSB (synchronization signal block) bursts):

- the index of the synchronization signal block (SSB Index) for beam sweeping of 64 bursts (FR2),

- or 1 most significant bit for the offset of the SSB block (K_{ssb}) and 2 bits not used in the specification of Release 16;

– 1 bit indicating the half-frame.

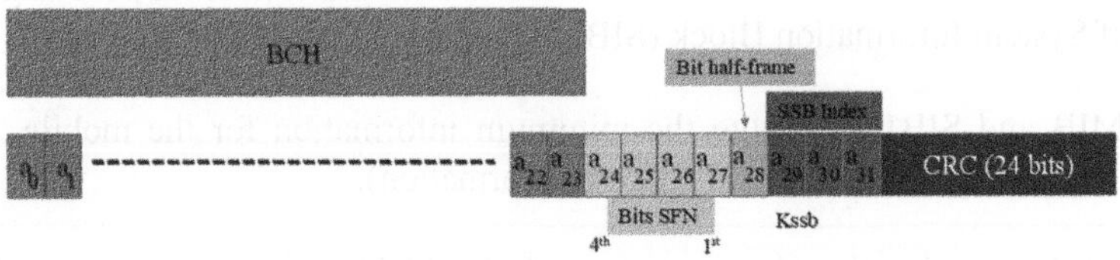

Figure 9.1. *PBCH physical channel. For a color version of this figure, see www.iste.co.uk/launay/5g.zip*

The SI-RNTI (System Information – Radio Network Temporary Identifier) is used to retrieve the description of the resource assigned to SIB messages in the PDSCH (Physical Downlink Shared Channel). This description is provided by the PDCCH (Physical Downlink Shared Channel).

9.1.1. *MIB message*

The MIB message is transmitted in the synchronization signal block.

The MIB provides the information needed by the mobile to read the PDCCH.

During this step, the information carried by the PDCCH only makes it possible to read the SIB1 message.

The MIB message contains the following parameters:

```
MasterInformationBlock
{
    systemFrameNumber
    subCarrierSpacingCommon
    ssb-Subcarrieroffset
    dmrs-TypeA-Position
    pdcch-ConfigSIB1
    cellBarred
    intraFreqReselection}

pdcch-ConfigSIB1
{
controlResourceSetZero ControlResourceSetZero,
searchSpaceZero SearchSpaceZero
}
```

systemFrameNumber: this parameter defines the 6 most significant bits of the frame number. The four least significant bits are acquired during decoding of the PBCH.

subCarrierSpacingCommon: this parameter indicates the spacing between sub-carriers concerning the SIB1 message and the spacing between sub-carriers of messages 2 and 4 sent by the gNB entity during the RACH procedure (see Figures 3.1 and 3.7). Two choices are possible, scs15or60 or scs30or120:

– in the FR1 band (Frequency Range 1), i.e. 15 kHz or 30 kHz;

– in the FR2 band, either 60 kHz or 120 kHz.

ssb-Subcarrieroffset: this 4-bit parameter concatenated with the most significant bit acquired during the decoding of the PBCH gives the value of the K_{ssb} parameter (see Figure 6.14).

If the value of the K_{ssb} parameter is greater than 23 in the FR1 band (Frequency Range 1) or greater than 11 in the FR2 band, then the BWP (BandWidth Partition) on which the mobile decodes the data does not contain a common search area *PDCCH_type0*. This means the BWP is not an initial partition.

dmrs-TypeA-Position: this 1-bit parameter indicates the position in the time domain of the first DM-RS (Demodulation Reference Signal) symbol for the PDSCH (Physical Downlink Shared Channel) and for the PUSCH (Physical Uplink Shared Channel). If *dmrs-TypeA-Position = 0*, the time position of the reference signal is on the 2nd symbol (pos2) and if *dmrs-TypeA-Position = 1*, the time position of the reference signal is on the 3rd symbol (pos3).

pdcch-ConfigSIB1: this 8-bit parameter is used to configure search space 0 for the initial BWP. This information allows the mobile to monitor the PDCCH in order to be aware of the presence of the SIB1 information message. If the BWP is not an initial band (not CORESET#0), then the *pdcch-ConfigSIB1* parameter indicates the frequency position of the SSB of the initial BWP of the cell.

cellBarred: this 1-bit parameter indicates whether the mobile can camp on the cell or not.

intraFreqReselection: if the cell is prohibited, this 1-bit parameter indicates whether the mobile can search for another cell on the same carrier band.

9.1.2. *SIB1 message*

The SIB1 message provides the information that allows the mobile to know whether it is authorized to access the cell and, if applicable, the information necessary to retrieve the other SIB messages.

The SIB1 message contains the following parameters:

```
SystemInformationBlockType1
    cellSelectionInfo
    cellAccessRelatedInfo
    ConnEstFailureControl
    SI-schedulingInfo
    servingCellConfigCommon
    ims-EmergencySupport
    ecallOver-IMS-Support
    ue-TimerAndConstants
    uac-BarringInfo
```

cellSelectionInfo: this parameter provides the minimum value expected to select the cell. The mobile selects the cell if the power measured from the Reference Signal Received Power (RSRP) is greater than the threshold.

cellAccessRelatedInfo: this parameter provides various pieces of information concerning access to the cell, such as the identities of the networks, MCC (Mobile Country Code) and MNC (Mobile Network Code), which share the cell, the number of the TACs (Tracking Area Codes) and cell identity.

ConnEstFailureControl: in the event of no response from the gNB during a radio connection request, the mobile increments a connection establishment failure counter. When this counter reaches a threshold, the power measured on the RSRP reference signal of the cell on which the mobile is camping is degraded. This parameter provides information on the number of failures before applying a penalty of the measure, and provides the value of the penalty as well as the duration.

SI-schedulingInfo: this parameter provides the information that defines the grouping in SI (System Information) messages SIB2 to SIB9 and the periodicity of each group.

servingCellConfigCommon: this parameter provides the configuration of the radio access on the uplink, downlink and for the UL band and the configuration of the TDD mode.

In the downlink direction, the *servingCellConfigCommon* parameter provides the configuration of the initial BWP tape partition, the spacing between subcarriers.

In the uplink direction, this parameter provides the information necessary to establish the initial access procedure in SA (StandAlone) mode.

ims-EmergencySupport: this parameter, introduced in Release 9, indicates whether the cell supports emergency calls for a mobile that cannot be attached to the AMF (Access and Mobility Management Function) entity. When the mobile is not registered, it is in a limited service state. The emergency call also requires the activation of the location service.

ecallOver-IMS-Support: the e-call system is an in-vehicle system allowing emergency calls and the transmission of additional MSD (Minimum Set of Data) data. The emergency e-call via IMS support requires the presence of a UICC card. This parameter, introduced in Release 14, indicates whether or not the cell accepts e-calls.

ue-TimerAndConstants: this parameter provides the value of different timers

– timer T300 relates to the procedure for establishing the RRC (Radio Resource Control) connection and indicates the mobile waiting time for the response to the RRC *SetupRequest* message;

– timer T301 starts after sending the RRC *ReEstablishmentRequest* message. The mobile goes into standby if it has not received a response when the timer expires;

– timer T310 starts if the mobile has received N310 desynchronization indications. The mobile goes into standby if it has not received N311 synchronization indications;

– timer T311 starts during the RRC connection re-establishment procedure. The mobile goes to sleep if it has not managed to find a cell to connect to;

– counter N311 is the maximum value of consecutive synchronizations with the main PCell;

– timer T319 starts when the RRC *ResumeRequest* message is transmitted. If one of the following answers is missing: RRC Resume, RRC Setup, RRC Release, RRC *ReleasewithsuspendConfig*, RRC *Rejectmessage*, the mobile goes into standby.

uac-barringinformation: this parameter makes it possible to provide an access control mechanism for the cell according to the access class stored in the SIM application.

The access classes are defined by the following values:

– Access Class 10 to initiate an emergency call;

– Access Class 11 for radio access from the home operator;

– Access Class 12 for security services;

– Access Class 13 for public services (water and gas distribution, etc.);

– Access Class 14 is reserved for emergency calls;

– Access Class 15 is intended for FUTs (Friendly User Tests).

uac-barringinformation contains the following information:

– *cellReservedForOperatorUse*: the cell can be reserved for users of the HPLMN/EPLMN operator whose mobile access identity is equal to 11, or to employees of the operator whose mobile access identity is equal to 15 and not authorized for mobiles with access identity 0, 1, 2, 12, 13 or 14;

– *cellReservedForOtherUse*: indicates whether or not the cell is reserved for another use.

9.1.3. *SIB2 message*

The SIB2 message provides the common information relating to the re-selection of intra-/inter-frequency and inter-RAT (Radio Access Technology) cells.

The SIB2 message contains the following parameters:

```
SystemInformationBlockType2
    cellReselectionInfoCommon
    cellReselectionServingFreqInfo
    intraFreqCellselectionInfo
    Q-OffsetRange
    SSB-MTC
    SS-RSSI-Measurement
    SSB-ToMeasuse
```

cellReselectionInfoCommon: this parameter provides the hysteresis value to be used for the selection of a new intra-/inter-frequency or inter-RAT cell and the number of resource blocks over which the power measurement must be averaged for intra-frequency re-selection.

cellReselectionServingFreqInfo: this parameter provides the conditions for the selection of either a new 5G cell with a different frequency (inter-frequency) from that of the server cell or a new 4G cell.

intraFreqCellReselectionInfo: this parameter provides the conditions for the selection of a new 5G cell of the same frequency (intra-frequency) as that of the server cell.

Q-OffsetRange: this parameter provides the offset value to apply for the selection of a cell or a beam.

SSB-MTC: this parameter provides the SMTC (SSB-based Measurement Timing Configuration) corresponding to the occasions of measurement of the SSB by the mobile.

SS-RSSI-Measurement: this parameter provides the configuration of the RSSI (NR carrier Received Signal Strength Indicator) measurement performed on the reference signals.

SSB-ToMeasure: this parameter provides the set of SSBs to be measured during the SMTC time. In the absence of this parameter, all SSBs must be measured.

9.1.4. *SIB3 message*

The SIB3 message provides the information which allows the mobile to select an intra-frequency neighboring cell. The SIB3 message contains the following parameters:

```
SystemInformationBlockType3
    intraFreqNeighCellList
    intraFreqBlackCellList
```

The two parameters are optional, the mobile having the ability to select a new intra-frequency cell without this information.

intraFreqNeighCellList: this parameter provides a list of neighboring PSCells identified by their PCI (Physical layer Cell Identity) and the minimum RSRP power level expected for the main cell and the additional SUL (Supplementary UpLink).

intraFreqBlackCellList: this setting provides a list of forbidden neighbor cells (up to 16 forbidden cell ranges).

9.1.5. *SIB4 message*

The SIB4 message provides information relating to neighboring 5G cells of different frequencies (inter-frequency) from that of the server cell and the re-direction criteria. The SIB4 message contains the following parameters:

```
SystemInformationBlockType4
    interFreqCarrierFreqList
```

interFreqCarrierFreqlList: this parameter provides the ARFCN frequencies of the downlink carriers and the common information relating to the neighboring inter-frequency cells (like the information provided by SIB2: cell re-selection, SMTC, etc.).

9.1.6. *SIB5 message*

The SIB5 message provides the information relating to the neighboring 4G cells and the re-selection criteria. The SIB5 message contains the following parameters:

```
SystemInformationBlockType5

    carrierFreqListEUTRA
    t-ReselectionEUTRA
    t-ReselectionEUTRA-SF
```

carrierFreqListEUTRA: this parameter provides information relating to neighboring cells operating in 4G mode and the re-selection criteria.

t-ReselectionEUTRA: this parameter provides a timeout value for the selection of the 4G E-UTRAN cell.

t-ReselectionUTRA-SF: this parameter provides the multiplicative factor applied to the *t-ReselectionEUTRA* parameter in order to take into account the mobility of the terminal (high or medium displacement).

9.1.7. *SIB6 message*

The SIB6 message provides, for the Japanese version, the brief ETWS (Earthquake and Tsunami Warning System) information. The SIB6 message contains the following parameters:

```
SystemInformationBlockType10
    messageIdentifier
    serialNumber
    warningType
```

messageIdentifier: this parameter provides the type of notification.

serialNumber: this setting provides a trace of changes from different notifications.

warningType: this parameter indicates whether the notification relates to an earthquake only, a tsunami only, or an earthquake and a tsunami.

9.1.8. *SIB7 message*

The SIB7 message provides more detailed information about ETWS alerts. The SIB7 message contains the following parameters:

```
SystemInformationBlockType11
    messageIdentifier
    serialNumber
    warningMessageSegmentType
    warningMessageSegmentNumber
    warningMessageSegment
    dataCodingScheme
```

messageIdentifier: this parameter provides the type of notification.

serialNumber: this setting provides a trace of changes to different notifications.

warningMessageSegmentType: this parameter indicates whether the *warningMessageSegment* parameter is the last segment constituting the message.

warningMessageSegmentNumber: this parameter indicates the segment number.

warningMessageSegment: this parameter contains the segment of the alert message.

dataCodingScheme: this parameter defines the syntax used for the alert message.

9.1.9. *SIB8 message*

The SIB8 message, introduced in Release 9, provides Commercial Mobile Alert System (CMAS) information for the North American version. The SIB8 message contains the same parameters as the SIB7 message:

```
SystemInformationBlockType12
    messageIdentifier
    serialNumber
    warningMessageSegmentType
    warningMessageSegmentNumber
    warningMessageSegment
    dataCodingScheme
```

9.1.10. *SIB9 message*

The SIB9 message carries the GPS information of the base station and the UTC (Coordinated Universal Time).

9.1.11. *Summary*

The minimum information is carried by the MIB and SIB1. Once the MIB is received, SIB1 contains the RMSI. Table 9.2 summarizes the different messages carried by each SIB.

Type	Content
MIB	Number of SFN frames, critical information for the reception of SIB1, flag on the authorized or unauthorized state of the cell and intra-frequency cell re-selection
SIB1	Cell selection information, the PLMN list, Cell identity, TAC, RAN code, flag for reserved cell, connection failure setting, sequencing information for other ISs, configuration of cell procedures (RACH, paging, SUL, etc.), TDD UL/DL mode configuration, VoIP emergency call support flag, timers values, access control information
SIB2	Cell re-selection information
SIB3	Information on the frequencies of the server cell and neighboring cells in the case of the re-selection procedure
SIB4	Information on the other NR carriers of the server cell and neighboring cells in the event of a re-selection procedure
SIB5	Information on the E-UTRA carriers of the server cell and neighboring cells in the event of a re-selection procedure
SIB6	ETWS primary notification
SIB7	ETWS secondary notification
SIB8	CMAS (Commercial Mobile Alert System) notification
SIB9	Base station GPS and UTC information

Table 9.2. *MIB and SIB information*

9.2. Connection management

9.2.1. *Paging*

The paging procedure consists of transmitting a notification to the mobile in the RRC_IDLE state or in the RRC_INACTIVE state to inform it of incoming data.

Table 9.3 provides the characteristics of the paging message support.

SRB	Mode RLC	Logical channel	Transport channel	Physical channel
No object	TM	PCCH	PCH	PDSCH
Paging				

Table 9.3. *The support of the message relating to paging*

The paging message is sent by the gNB entity and contains the following information elements:

```
Paging
   pagingRecordList
```

pagingRecordList: this information element indicates the identity of the mobile recipient of the notification, ng-5G-S-TMSI (Shortened Temporary Mobile Subscriber Identity), I-RNTI or IMSI (International Mobile Subscriber Identity), as well as the origin of the notification in PS (Packet Service) mode on a 3GPP or non-3GPP network.

9.2.2. *Connection establishment*

The connection establishment procedure follows the random access procedure and allows the transition from the RRC_IDLE state to the RRC_CONNECTED state and the establishment of the SRB1 (Signaling Radio Bearer).

Table 9.4 provides the characteristics of the message support relating to the establishment of the connection.

SRB	RLC mode	Logical channel	Transport channel	Physical channel
SRB0	TM	CCCH	UL-SCH	PUSCH
		SetupRequest		
SRB0	TM	CCCH	DL-SCH	PDSCH
		SetupReject		
SRB1	AM	DCCH	UL-SCH	PUSCH
		SetupComplete		

Table 9.4. *Support for messages relating to the connection establishment*

The *SetupRequest* message is transmitted by the mobile to initiate the connection establishment procedure. The message contains the following pieces of information:

```
RRCSetupRequest
   ue-Identity
   establishmentCause
```

ue-Identity: this element of information indicates the identity of the mobile, the 5G-S-TMSI identity if the mobile is attached or a 39-bit random value otherwise.

establishmentCause: this element of information indicates the cause of the connection establishment as:

– emergency call;

– priority call;

– mobile terminated call MT;

– mobile originating call (MO-Data MO-VoiceCall, MO-VideoCall, MO-SMS);

– mobile originating signaling (*MO-Signaling*);

– priority multimedia service (*mps-PriorityAccess*);

– critical service mission (*mcs-PriorityAccess*).

The RRC *Setup* message is the response from the gNB entity to accept the request to establish the connection.

The RRC *Setup* message is transmitted to establish the SRB1 bearer and carries the information for the configuration of the radio bearer of the master cell.

The RRC *Setup* message is used to configure the parameters of the PHY, MAC, RLC, PDCP and SDAP layers, and contains the following information elements:

```
RRCCSetup
    rrc-TransactionIdentifier
    radioBearingConfig
    CellGroupConfig
```

rrc-TransactionIdentifier: this element of information provides the transaction ID.

radioBearerConfig: this element of information provides the characteristics of the SRB1 or SRB3 media and the DRB media. The configuration changes the parameters of the PDCP sublayer.

cellGroupConfig: this element of information provides the RLC, MAC and PHY parameters for the configuration of a secondary cell. The secondary cell corresponding to carrier aggregation or dual connectivity (PSCell).

The RRC *Reject* message constitutes the response of the gNB entity to reject the request to establish the connection and contains the following information elements:

```
RRCReject
    waitTime
    extendedWaitTime
```

waitTime: this element of information indicates the waiting time which is between 1 and 16 seconds.

extendedWaitTime: this element of information, introduced in Release 10, extends the wait time to 1800 seconds.

The RRC *SetupComplete* message constitutes the response from the mobile to the RRC *Setup* message to confirm the establishment of the connection and contains the following information elements:

```
RRCSetupComplete
    rrc-TransactionIdentifier
    selectedPLMN-Identity
    registeredAMF
    guami-Type
    s-nssai-List
    dedicatedNAS-message
```

selectedPLMN-Identity: this element of information defines the identity of the MCC (Mobile Country Code) and MN C (Mobile Network Code) on which the mobile wishes to connect, in the case where the cell is shared between several networks.

registeredAMF: this element of information indicates the GUAMI (Globally Unique AMF Identity) of the AMF entity on which the mobile is attached.

Guami-Type: this element indicates whether the GUAMI is native, i.e. derived from 5G-GUTI or obtained by correspondence from the GUTI EPS.

s-nssai-List: this element indicates the list of S-NSSAIs supported by the AMF.

dedicatedNAS-message: this element of information indicates that the RRC message carries an NAS (Non-Access Stratum).

9.2.3. *Activation of security*

The security activation procedure starts when the gNB entity receives a request from the AMF to establish the context for the mobile.

The procedure for activating the security of the radio interface allows the implementation of the following operations:

– encryption and integrity control (if enabled) of the data on the DRB media;

– encryption and data integrity control of the SRB1 and SRB2 media.

Table 9.5 provides the characteristics of the message support relating to the activation of the security of the access stratum.

SRB	RLC mode	Logical channel	Transport channel	Physical channel
SRB1	AM	DCCH	DL-SCH	PDSCH
SecurityModeCommand				
SRB1	AM	DCCH	UL-SCH	PUSCH
SecurityModeComplete *SecurityModeFailure*				

Table 9.5. *Support for messages relating to the activation of security*

The *SecurityModeCommand* message is transmitted by the gNB entity to define the algorithms used for the encryption of SRB and DRB media and for the integrity check of the SRB media.

Upon receipt of the *SecurityModeCommand* message, the mobile performs the following actions:

– derives the K_{gNB} key from the K_{AMF} key;

– derives the K_{RRCint} integrity key and the K_{upint} key associated with the integrity protection algorithm;

– checks the integrity protection of the *SecurityModeCommand* message;

– starts the integrity protection of the SRB media.

The *SecurityModeComplete* message is transmitted by the mobile to confirm the activation of security. Encryption is operational after transmitting this message. The *SecurityModeComplete* message is not encrypted but integrity protected.

The *SecurityModeFailure* message is transmitted by the mobile to indicate the failure to activate security.

9.2.4. Connection reconfiguration

Connection reconfiguration messages are used to modify an existing RRC connection. The procedure concerns the following operations:

– the establishment, modification and release of DRB media;

– the establishment, modification and release of radioelectric measurements;

– the establishment, modification and release of SCell secondary cells;

– triggering of the handover.

Table 9.6 provides the characteristics of the message support relating to the reconfiguration of the connection.

SRB	RLC mode	Logical channel	Transport Channel	Physical Channel
SRB1	AM	DCCH	DL-SCH	PDSCH
Reconfiguration				
SRB1	AM	DCCH	UL-SCH	PUSCH
ReconfigurationComplete				

Table 9.6. *Support for messages relating to the reconfiguration of the connection*

The *Reconfiguration* message is transmitted by the gNB entity and allows us to configure the parameters of the PHY, MAC, RLC, PDCP and SDAP layers.

The *Reconfiguration* message contains the following pieces of information:

```
RRCReconfiguration
    rrc-TransactionIdentifier
    measConfig
    radiobearerconfig
    SecondaryCellGroup
```

measConfig: this element of information defines the measurements to be performed on neighboring cells 5G (intra-frequency and inter-frequency) and 4G (inter-RAT).

radiobearerconfig: this element of information is used for the establishment, modification and release of the radio medium and to modify the configuration of the PDCP layer and the dedicated physical channels.

SecondaryCellGroup: this element of information indicates the list of SCells to be removed or added by configuring the parameters of the RLC, MAC and physical layer.

The *ReconfigurationComplete* message is transmitted by the mobile to confirm the reconfiguration of the connection. The handover procedure is finalized upon receipt of this message. The dual connectivity procedure is established even in the absence of this message.

9.2.5. *Connection re-establishment*

The connection re-establishment procedure allows the operation of the SRB1 support to resume, the reactivation of security and the configuration of the PCell.

Table 9.7 provides the characteristics of the connection re-establishment message support.

SRB	RLC mode	Logical channel	Transport channel	Physical channel
SRB0	TM	CCCH	UL-SCH	PUSCH
ReestablishmentRequest				
SRB0	TM	CCCH	DL-SCH	PDSCH
Reestablishment *ReestablishmentReject*				
SRB1	AM	DCCH	UL-SCH	PUSCH
ReestablishmentComplete				

Table 9.7. *Support for messages relating to connection re-establishment*

The *ReestablishmentRequest* message is transmitted by the mobile and contains the following information elements:

```
RRCConnectionReestablishmentRequest
    ue-Identity
    reestablishmentCause
```

ue-Identity: this element of information contains the identity of the mobile C-RNTI (Cell Radio Network Temporary Identity) and the physical identity of the PCI (Physical layer Cell Identity) cell to which the mobile was connected before the failure.

reestablishmentCause: this information element indicates the cause (handover, reconfiguration) of the triggering of the connection re-establishment.

Upon receipt of the RRC *ReestablishmentRequest* message, the gNB responds with the *RRCSetup* or *RRCReestablishment* message. The *Reestablishment* message is sent by the gNB entity and contains the following pieces of information:

```
RRCReestablishment
    radioResourceConfigDedicated
    nextHopChainingCount
```

radioResourceConfigDedicated: this element of information provides the characteristics of the radio carriers to be restored.

nextHopChainingCount: this piece of information provides the parameter used to renew the encryption and integrity control keys.

The *ReestablishmentReject* message is transmitted by the gNB to indicate the rejection of the request to re-establish the connection.

The *ReestablishmentComplete* message is transmitted by the mobile to confirm the re-establishment of the connection.

9.2.6. *Connection release*

The connection release procedure allows the radio resources to be released. The access network initiates the resource release connection, the mobile changes from the RRC_CONNECTED state to the RRC_IDLE state or to the RRC_INACTIVE state to suspend the connection.

Table 9.8 provides the characteristics of the connection release message bearer.

SRB	RLC mode	Logical channel	Transport channel	Physical channel
SRB1	AM	DCCH	DL-SCH	PDSCH
Release				

Table 9.8. *Support for messages relating to the connection release*

The *Release* message is transmitted by the gNB entity and contains the following information elements:

```
RRCRelease
    rrc-TransactionIdentifier
    redirectedCarrierInfo
    cellReselectionPriorities
    suspendConfig
    idleModeMobilityControlInfo
```

redirectedCarrierInfo: this element of information allows the mobile to be redirected on a 4G radio channel.

cellReselectionInfoPriorities: this element of information provides the priority levels for the selection of new cells.

suspendconfig: this element of information is used to suspend the configuration of the terminal. The context is saved, and the mobile goes in the to RRC_INACTIVE state.

9.3. Measurement configuration

In standby mode, the measurement configuration is broadcast by the gNb base station in the SIB 1 message (Broadcast Control Channel System Information Block 1).

When switching to connected mode, the configuration of the measurements to be performed by the mobile is triggered by the gNB entity in the *RRCReconfiguration*, *RRCResume* messages. The information is stored in the *MeasConfig* measurement configuration and contains the following parameters:

– the object that identifies the radio channel;

– the event that triggers the measurement report:

– the measurement format: the quantity of information to be returned to the base station and the associated information (e.g. several cells, several RATs);

– the criteria that trigger the measurements (periodic or by event);

– the combination of objects and events by measurement identity;

– the quality of the configuration and, in particular, the filtering parameter;

– the periodicity of measurements when no upward or downward communication is planned.

Since the mobile does not generally have several radio receivers, the inter-frequency and inter-RAT measurement must be carried out in the intervals provided in the frame.

The measurements made on the server cell and neighboring cells are used for cell selection and handover.

Intra-frequency measurement, essential to ensure mobility, is configured during the radio connection.

Inter-frequency and inter-RAT (Radio Access Technology) measurement can also be configured during connection.

The measurements made on the eNB master base station and the en-gNB secondary base station are used for dual connectivity.

9.3.1. *Measurement objects*

The measurements carried out on the radioelectric channels of neighboring cells are of different types:

– intra-frequency: the frequency of the neighboring cell is identical to that of the server cell;

– inter-frequency: the frequency of the neighboring cell is different from the server cell;

– inter-RAT E-UTRAN (Evolved Universal Terrestrial Radio Access Network).

The measurements performed on the secondary cell or neighboring cells concern the following values:

– for the 5G cell, the SSB-RSRP power level measured at the level of the SSB, the CSI-RSRP power level and the RSRQ quality level measured on the reference signal;

– for the 4G cell, the RSRP level and the RSRQ level measured on the reference signal.

9.3.2. *The measurement events*

9.3.2.1. *The A1 event*

The A1 event occurs when the measurement of the signal received from the serving cell becomes greater than that of a threshold:

$$Ms - Hys > Thresh$$

where:

– *Ms* is the measurement of the signal received on the server cell;

– *Hys* is the hysteresis parameter relating to this event;

– *Thresh* is the threshold relating to this event.

The A1 event is stopped when the measurement of the signal received from the serving cell becomes less than that of a threshold:

$$Ms + Hys < Thresh.$$

Example of use: the A1 event can be used to cancel a handover procedure.

9.3.2.2. *The A2 event*

The A2 event occurs when the measurement of the signal received from the serving cell falls below that of a threshold:

$$Ms + Hys < Thresh.$$

The A2 event is stopped when the measurement of the signal received from the serving cell becomes greater than that of a threshold:

$$Ms - Hys > Thresh.$$

Example of use: the A2 event can be used to trigger a mobility procedure when the mobile is in a cell device.

9.3.2.3. *The A3 event*

The A3 event occurs when the measurement of the signal received from the neighboring cell is greater than that of the SpCell (Special Cell):

$$Mn + Ofn + Ocn - Hys > Mp + Ofp + Ocp + Off$$

where:

– Mn is the measurement of the signal received from the neighboring cell;

– Ofn is the specific offset related to the frequency of the neighboring cell;

– Ocn is the specific offset related to the neighboring cell;

– Mp is the measurement of the level received on the primary channel SpCell of the serving cell. In the case of dual connectivity, the SpCell channel corresponds to the PCell channel of the MCG (Master Cell Group) or to the PSCell channel of the SCG (Secondary Cell Group), depending on whether the MAC entity is associated with the MCG or SCG;

– Ofp is the specific offset linked to the frequency of the serving cell;

– Ocp is the specific offset linked to the serving cell;

– Off is the specific offset relative to this event.

The A3 event is stopped when the measurement of the signal received from the neighboring cell becomes lower than that of the server cell:

$$Mn + Ofn + Ocn + Hys < Mp + Ofp + Ocp + Off$$

Example of use: the A3 event can be used for an intra- or inter-frequency handover procedure.

9.3.2.4. *The A4 event*

The A4 event occurs when the measurement of the signal received from the neighboring cell becomes greater than that of a threshold:

$$Mn + Ofn + Ocn - Hys > Thres.$$

The A4 event is stopped when the measurement of the signal received from the neighboring cell becomes less than that of a threshold:

$$Mn + Ofn + Ocn + Hys < Thres.$$

Example of use: the A4 event can be used to trigger a handover procedure when the main cell is overloaded.

9.3.2.5. *The A5 event*

The A5 event occurs when the measurement of the signal received from the SpCell becomes less than that of a first threshold and the measurement received from the neighboring cell becomes greater than that of a second threshold:

$$Mp + Hys < Thresh;$$

$$Mn + Ofn + Ocn - Hys > Thres2.$$

The A5 event is stopped when the measurement of the signal received from the server cell becomes greater than that of a first threshold and the measurement received from the neighboring cell becomes less than that of a second threshold:

$$Mp - Hys > Thresh;$$

$$Mn + Ofn + Ocn + Hys < Thres2.$$

Example of use: the A5 event can be used to trigger an intra- or inter-frequency handover procedure.

9.3.2.6. *The A6 event*

The A6 event occurs when the measurement of the signal received from the neighboring cell becomes greater than that of the signal received from the secondary channel of the serving cell (SCell):

$$Mn + Ocn - Hys > Ms + Ocs + Off$$

where Ocs is the specific offset linked to the secondary channel of the serving cell. The offset can be positive or negative.

The A6 event is stopped when the measurement of the signal received from the neighboring cell becomes lower than that of the signal received from the secondary channel of the serving cell (SCell):

$$Mn + Ocn + Hys < Ms + Ocs + Off.$$

9.3.2.7. *The B1 event*

The B1 event occurs when the measurement of the signal received from the neighboring inter-RAT (4G) cell with an inter-RAT offset ofn and ocn becomes greater than that of a threshold:

$$Mn + Ofn + Ocn - Hys > Thresh.$$

The B1 event is stopped when the measurement of the signal received from the neighboring inter-RAT (4G) cell falls below that of a threshold within one offset:

$$Mn + Ofn + Ocn + Hys < Thresh.$$

Event B1 is used for the handover procedure.

The B1 event is stopped when the measurement of the signal received from the neighboring inter-RAT (4G) cell falls below that of the threshold.

9.3.2.8. *The B2 event*

The B2 event occurs when the measurement of the signal received from the PCell is below a threshold and the 4G inter-RAT measurements become greater than a second threshold:

$$Mp + Hys < Thresh;$$

$$Mn + Ofn + Ocn - Hys > Thres2.$$

The B2 event is stopped when the measurement of the signal received from the PCell is less than a threshold and the inter-RAT 4G measurements become greater than a second threshold:

$$Mp - Hys > Thresh;$$

$$Mn + Ofn + Ocn + Hys < Thres2.$$

9.3.3. *The filtering of the measurement*

The filtering of the measurement consists of weighting the value communicated by the physical layer.

The filtering of each type of measurement is obtained by applying the following relationship:

$$F_n = (1-a).F_{n-1} + a.M_n$$

where:

– F_n is the updated value to measure;

– F_{n-1} is the previous measure, F_0 is the first measure, M_1 is given from the physical layer;

– $a = \left(\frac{1}{2}\right)^{\frac{k}{4}}$, in which k is the filter coefficient.

The measurements are reported by the UE (User Equipment) either periodically or when the UE is triggered by an event. Once the event has occurred, the measurements are reported once or periodically.

9.4. References

All standards can be downloaded from the ETSI website: https://www.etsi.org/standards.

3GPP 23.501
System architecture for the 5G System (5GS);
Version 16.6.0 Release 16

3GPP 36.304
LTE - Evolved Universal Terrestrial Radio Access (E-UTRA);
User Equipment (UE) procedures in idle mode;
Version 9.9.0 Release 9

3GPP 38.104
NR - Base Station (BS) radio transmission and reception;
Version 15.2.0 Release 15

3GPP 38.211
NR - Physical channels and modulation;
Version 16-2.0 Release 16

3GPP 38.212
NR - Multiplexing and channel coding;
Version 16-2.0 Release 16

3GPP 38.213
NR - *Physical layer procedure for control;*
Version 16-2.0 Release 16

3GPP 38.300
NR - NR and NG-RAN Overall description; Stage-2
Version 16-2.0 Release 16

3GPP 38.331
NR - Radio Resource Control (RRC); Protocol specification;
Version 16-2.0 Release 16

38.508
NR, 5GS -User Equipment (UE) conformance specification; Part 1: Common test environment;
Version 15.2.0 Release 15

5G Tools for RF Wireless (2021). 5G Tools for RF Wireless: Tools from engineers for engineers [Online]. Available at: https://5g-tools.com/5g-nr-gscn-calculator/.

Index

Q, R, S

U, V, X

Other titles from

iSTE

in

Networks and Telecommunications

2020

GONTRAND Christian
Digital Communication Techniques

PUJOLLE Guy
Software Networks: Virtualization, SDN, 5G and Security (2nd edition revised and updated)
(Advanced Network Set – Volume 1)

2019

LAUNAY Frédéric, PEREZ André,
LTE Advanced Pro: Towards the 5G Mobile Network
Harmonic Concept and Applications

TOUNSI Wiem
Cyber-Vigilance and Digital Trust: Cyber Security in the Era of Cloud Computing and IoT

2018

ANDIA Gianfranco, DURO Yvan, TEDJINI Smail
Non-linearities in Passive RFID Systems: Third Harmonic Concept and Applications

BOUILLARD Anne, BOYER Marc, LE CORRONC Euriell
Deterministic Network Calculus: From Theory to Practical Implementation

LAUNAY Frédéric, PEREZ André
LTE Advanced Pro: Towards the 5G Mobile Network

PEREZ André
Wi-Fi Integration to the 4G Mobile Network

2017

BENSLAMA Malek, BENSLAMA Achour, ARIS Skander
Quantum Communications in New Telecommunications Systems

HILT Benoit, BERBINEAU Marion, VINEL Alexey, PIROVANO Alain
Networking Simulation for Intelligent Transportation Systems: High Mobile Wireless Nodes

LESAS Anne-Marie, MIRANDA Serge
The Art and Science of NFC Programming
(Intellectual Technologies Set – Volume 3)

2016

AL AGHA Khaldoun, PUJOLLE Guy, ALI-YAHIYA Tara
Mobile and Wireless Networks
(Advanced Network Set – Volume 2)

BATTU Daniel
Communication Networks Economy

BENSLAMA Malek, BATATIA Hadj, MESSAI Abderraouf
Transitions from Digital Communications to Quantum Communications: Concepts and Prospects

CHIASSERINI Carla Fabiana, GRIBAUDO Marco, MANINI Daniele
Analytical Modeling of Wireless Communication Systems (Stochastic Models in Computer Science and Telecommunication Networks Set – Volume 1)

EL FALLAH SEGHROUCHNI Amal, ISHIKAWA Fuyuki, HÉRAULT Laurent, TOKUDA Hideyuki
Enablers for Smart Cities

PEREZ André
VoLTE and ViLTE

2015

BENSLAMA Malek, BATATIA Hadj, BOUCENNA Mohamed Lamine
Ad Hoc Networks Telecommunications and Game Theory

BENSLAMA Malek, KIAMOUCHE Wassila, BATATIA Hadj
Connections Management Strategies in Satellite Cellular Networks

BERTHOU Pascal, BAUDOIN Cédric, GAYRAUD Thierry, GINESTE Matthieu
Satellite and Terrestrial Hybrid Networks

CUADRA-SANCHEZ Antonio, ARACIL Javier
Traffic Anomaly Detection

LE RUYET Didier, PISCHELLA Mylène
Digital Communications 1: Source and Channel Coding

PEREZ André
LTE and LTE Advanced: 4G Network Radio Interface

PISCHELLA Mylène, LE RUYET Didier
Digital Communications 2: Digital Modulations

2014

ANJUM Bushra, PERROS Harry
Bandwidth Allocation for Video under Quality of Service Constraints

BATTU Daniel
New Telecom Networks: Enterprises and Security

BEN MAHMOUD Mohamed Slim, GUERBER Christophe, LARRIEU Nicolas, PIROVANO Alain, RADZIK José
Aeronautical Air–Ground Data Link Communications

BITAM Salim, MELLOUK Abdelhamid
Bio-inspired Routing Protocols for Vehicular Ad-Hoc Networks

CAMPISTA Miguel Elias Mitre, RUBINSTEIN Marcelo Gonçalves
Advanced Routing Protocols for Wireless Networks

CHETTO Maryline
Real-time Systems Scheduling 1: Fundamentals
Real-time Systems Scheduling 2: Focuses

EXPOSITO Ernesto, DIOP Codé
Smart SOA Platforms in Cloud Computing Architectures

MELLOUK Abdelhamid, CUADRA-SANCHEZ Antonio
Quality of Experience Engineering for Customer Added Value Services

OTEAFY Sharief M.A., HASSANEIN Hossam S.
Dynamic Wireless Sensor Networks

PEREZ André
Network Security

PERRET Etienne
Radio Frequency Identification and Sensors: From RFID to Chipless RFID

REMY Jean-Gabriel, LETAMENDIA Charlotte
LTE Standards
LTE Services

TANWIR Savera, PERROS Harry
VBR Video Traffic Models

VAN METER Rodney
Quantum Networking

XIONG Kaiqi
Resource Optimization and Security for Cloud Services

2013

ASSING Dominique, CALÉ Stéphane
Mobile Access Safety: Beyond BYOD

BEN MAHMOUD Mohamed Slim, LARRIEU Nicolas, PIROVANO Alain
Risk Propagation Assessment for Network Security: Application to Airport Communication Network Design

BERTIN Emmanuel, CRESPI Noël
Architecture and Governance for Communication Services

BEYLOT André-Luc, LABIOD Houda
Vehicular Networks: Models and Algorithms

BRITO Gabriel M., VELLOSO Pedro Braconnot, MORAES Igor M.
Information-Centric Networks: A New Paradigm for the Internet

DEUFF Dominique, COSQUER Mathilde
User-Centered Agile Method

DUARTE Otto Carlos, PUJOLLE Guy
Virtual Networks: Pluralistic Approach for the Next Generation of Internet

FOWLER Scott A., MELLOUK Abdelhamid, YAMADA Naomi
LTE-Advanced DRX Mechanism for Power Saving

JOBERT Sébastien *et al.*
Synchronous Ethernet and IEEE 1588 in Telecoms: Next Generation Synchronization Networks

MELLOUK Abdelhamid, HOCEINI Said, TRAN Hai Anh
Quality-of-Experience for Multimedia: Application to Content Delivery Network Architecture

NAIT-SIDI-MOH Ahmed, BAKHOUYA Mohamed, GABER Jaafar, WACK Maxime
Geopositioning and Mobility

PEREZ André
Voice over LTE: EPS and IMS Networks

2012

AL AGHA Khaldoun
Network Coding

BOUCHET Olivier
Wireless Optical Communications

DECREUSEFOND Laurent, MOYAL Pascal
Stochastic Modeling and Analysis of Telecoms Networks

DUFOUR Jean-Yves
Intelligent Video Surveillance Systems

EXPOSITO Ernesto
Advanced Transport Protocols: Designing the Next Generation

JUMIRA Oswald, ZEADALLY Sherali
Energy Efficiency in Wireless Networks

KRIEF Francine
Green Networking

PEREZ André
Mobile Networks Architecture

2011

BONALD Thomas, FEUILLET Mathieu
Network Performance Analysis

CARBOU Romain, DIAZ Michel, EXPOSITO Ernesto, ROMAN Rodrigo
Digital Home Networking

CHABANNE Hervé, URIEN Pascal, SUSINI Jean-Ferdinand
RFID and the Internet of Things

GARDUNO David, DIAZ Michel
Communicating Systems with UML 2: Modeling and Analysis of Network Protocols

LAHEURTE Jean-Marc
Compact Antennas for Wireless Communications and Terminals: Theory and Design

PALICOT Jacques
Radio Engineering: From Software Radio to Cognitive Radio

PEREZ André
IP, Ethernet and MPLS Networks: Resource and Fault Management

RÉMY Jean-Gabriel, LETAMENDIA Charlotte
Home Area Networks and IPTV

TOUTAIN Laurent, MINABURO Ana
Local Networks and the Internet: From Protocols to Interconnection

2010

CHAOUCHI Hakima
The Internet of Things

FRIKHA Mounir
Ad Hoc Networks: Routing, QoS and Optimization

KRIEF Francine
Communicating Embedded Systems / Network Applications

2009

CHAOUCHI Hakima, MAKNAVICIUS Maryline
Wireless and Mobile Network Security

VIVIER Emmanuelle
Radio Resources Management in WiMAX

2008

CHADUC Jean-Marc, POGOREL Gérard
The Radio Spectrum

GAÏTI Dominique
Autonomic Networks

LABIOD Houda
Wireless Ad Hoc and Sensor Networks

LECOY Pierre
Fiber-optic Communications

MELLOUK Abdelhamid
End-to-End Quality of Service Engineering in Next Generation Heterogeneous Networks

PAGANI Pascal *et al.*
Ultra-wideband Radio Propagation Channel

2007

BENSLIMANE Abderrahim
Multimedia Multicast on the Internet

PUJOLLE Guy
Management, Control and Evolution of IP Networks

SANCHEZ Javier, THIOUNE Mamadou
UMTS

VIVIER Guillaume
Reconfigurable Mobile Radio Systems

Printed and bound by CPI Group (UK) Ltd, Croydon, CR0 4YY
15/09/2026
14962359-0006